Your Superstar Brain

KAJA NORDENGEN

Your Superstar Brain

Unlocking the Secrets of the Human Mind

piatkus

PIATKUS

First published in Norway in 2016 by Kagge Forlag AS
First published in Great Britain in 2018 by Piatkus

1 3 5 7 9 10 8 6 4 2

This translation has been published with the financial support of NORLA

A CIP catalogue record for this book is available from the British Library.

ISBN 978-0-349-41722-6

Illustrations by Guro Nordengen
Photograph on page 185 © Geir Mogen; edited by Birte Nordengen

Typeset in Aldus by M Rules
Printed and bound in Great Britain by
Clays Ltd, St Ives plc

Papers used by Piatkus are from well-managed forests and other responsible sources.

Piatkus
An imprint of
Little, Brown Book Group
Carmelite House
50 Victoria Embankment
London EC4Y 0DZ

An Hachette UK Company
www.hachette.co.uk

About the author

Kaja Nordengen is a brain researcher and physician specialising in neurology at Akershus University Hospital. She has a PhD in neuroscience and also teaches at the University of Oslo.

Contents

FOREWORD

*by May-Britt Moser**

The brain is the most wondrous, complex and mysterious organ we know. As a psychology student in the 1980s, I was taught that the cause of autism in children was an emotionally cold mother. Today we know better. Autism is due to a developmental change in the brain, which brings a multitude of factors into the picture.

For me, this memory of my student days serves as a reminder of how quickly our knowledge has developed within the field of brain research. We should rejoice at the progress being made, but we should also remain humble when it comes to recognising the role that modern technology plays in making this new knowledge possible. Many of the big research questions of our day are the same ones that people have been asking for centuries. However, thanks to the development of ground-breaking research tools and methods, we are now able to look for the answers to these questions in the brain itself. We are standing on the threshold of a knowledge revolution in terms of the

---

* May-Britt Moser is a psychologist, brain researcher and professor of neuroscience at the Norwegian University of Science and Technology (NTNU). In 2014 she received the Nobel Prize in Medicine, along with Edvard Moser and John O'Keefe.

brain and its interaction with the body, genetics and the environment.

But it's not enough to collect research data in laboratories and share the results internationally with colleagues. The knowledge has to be conveyed beyond professional settings to society at large, where it may be incorporated into people's lives and translated into insight and understanding. To understand how our brain functions and how it participates in all of the body's processes is to understand our capabilities and who we are as human beings.

Greater knowledge also leads to superior means of evaluation and treatment when something goes wrong in the brain. We now understand the importance of separating the symptoms of a brain disorder from the person's character and personality. We know that the symptoms are due to a failure in the system. Armed with greater knowledge about how the healthy brain functions, researchers can proceed to look for where in its processes the failure has occurred and how it may be repaired. This insight provides the basis for largesse and willingness to adapt, which are essential for everyone to have a place in society.

In this book, Kaja Nordengen presents an accessible introduction to the most recent research into the organisation, mechanisms and functions of the brain. Writing in an engaging manner, she interweaves the results of this research with anecdotes from her own life. By anchoring theory in concrete experiences that are part of our shared world, she not only conveys facts but stimulates curiosity. The playfulness in the way she presents the material awakens the sort of enthusiasm that drives both the questioning child and the experienced adult. The warmth of Kaja's narrative voice is sure to stay with every reader long after they have closed her book.

The excellent illustrations were drawn by the author's younger sister, Guro Nordengen. Once again, I suspect that readers will remember these clear, simple drawings for many years to come, which is not usually the case with sophisticated 3D graphics. They mirror the text by encouraging further contemplation and general understanding rather than attempting to present every minute detail.

I would like to thank Kaja Nordengen for daring to take on this bold and ambitious project. Her tireless efforts mean that some of the most important advances in brain research are now accessible to a broad cross section of readers, both young and old.

Your Superstar Brain

Introduction

You Are Your Brain

When the ancient Egyptians embalmed their dead rulers in preparation for the afterlife, the heart was carefully removed before being placed back inside the body. Meanwhile, the brain was simply discarded. A stick was poked up the nose, the brain matter was whipped into mush, then it was sucked out. The brain became trash. It would be a long time before humans understood that we are who we are because of our brains.

Even before recorded history, the brain was sometimes linked to functions such as movement and thinking. Yet it would be several thousand years before people generally accepted that 'I' is situated within the brain. Aristotle and several other great classical thinkers, for example, believed that the soul was to be found in the heart. It was only in the mid-seventeenth century that the French philosopher René Descartes proposed a different hypothesis.

Almost everything on each side of the central line that divides the brain's two halves has a corresponding feature on the opposite side. For instance, we have a left and a right frontal lobe. Yet Descartes noticed that the pineal gland was located in the very centre, so he identified this as the seat of the human soul. But it wasn't quite that simple. In 1887, the Arctic explorer – and Norway's first brain researcher – Fridtjof Nansen correctly postulated that intelligence lies in the brain's numerous neural synapses. Since his day, we have learned that joy, love, contempt, memory, knowledge, musical taste and every other human preference are also located in these neural synapses.

Since every trait that makes up 'I' exists in the brain, clearly you could not have become you without your brain. Our laws also recognise that the brain is the principal determining factor for what constitutes life: if you are 'brain dead', then you are dead. Provided that permission has been granted, your organs can then be removed and transplanted to save someone else's life. There are few organs that we can manage without, yet the heart, liver, lungs, kidneys and pancreas are all replaceable, as long as a suitable donor can be found. But no one has yet attempted to transplant a human brain into another body.

Even when the technical challenges of doing so have been overcome, significant ethical dilemmas will remain. If someone who is brain dead receives a new brain, the person associated with that particular body will no longer be the same 'I'. She may look like your daughter; but if she has someone else's brain, is she really your daughter? She will have a completely different consciousness – alternative thoughts and dreams. We cannot replace the brain without also replacing the person. In other words, the brain is our only irreplaceable organ.

In this book, we will explore the mysteries of the brain – everything from what happens when we fall in love to where to find the 'I'. Lots of interesting questions arise when we talk about the brain. Who are you? What makes you who you are? What is personality? What is free will? Where do our thoughts originate? We already have some clear answers ... or at least some clear indications from patient histories and advances in brain research. Nevertheless, many mysteries remain, so we must hope that further research and bright minds will find more answers in the years to come. After all, the brain is the only organ that can research itself.

Language, culture and lifestyle all involve memory and the brain's ability to detect and interpret patterns. The brain makes us who we are, and it's the reason why sports, art and music exist. Your brain is a superstar.

Chapter 1

Thought (R)evolution

The human brain's wrinkly surface, which is reminiscent of the outside of a walnut, is called the cerebral cortex. It is packed full of neurons, and marked a revolution in the history of evolution. The bigger an animal's cerebral cortex, the greater its chances of having high intelligence.

Half a billion years ago, only the so-called reptile brain existed. This is now known as the rhombencephalon. The early mammalian brain – what we call the limbic system – appeared some 250 million years later, while the cerebral cortex developed about 200 million years ago. However, what we consider the human brain did not appear until 200,000 years ago – almost the equivalent of yesterday in evolutionary terms.

The reptile brain

The human race's large cerebral cortex helped us to survive the last ice age as it allowed us to adapt to the changing environment. By contrast, 65 million years ago, when a meteor strike generated major climatic change, the dinosaurs were not very well equipped to cope. For instance, an adult stegosaur weighed in at a whopping five tons but had a brain that weighed only eighty grams (about the size of a lemon). Moreover, this mini-brain lacked a cerebral cortex. When you know that, it's hardly surprising that you find stegosaurs only in movies and museums today.

However, while our cerebral cortex makes us the most intelligent species on the planet, we wouldn't have made it this far without the deeper parts of our brain. Deepest of all – and therefore most fundamental to our existence – is the reptile brain, which consists of the brainstem and the cerebellum. The brainstem is the perfect caretaker: it ensures that everything keeps functioning without any need for conscious thought. Its neurons regulate our breathing, cardiac rhythm and sleep. They never rest, whether we are awake or asleep. Meanwhile, the cerebellum, which sits behind the brainstem, regulates our movements. Therefore, we become uncoordinated and unsteady if it is affected by alcohol.

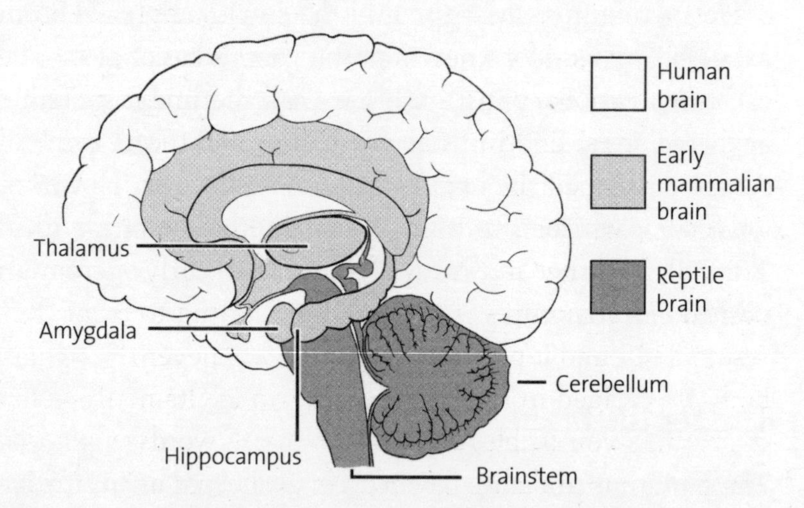

Figure 1: The right half of a human brain, seen from the middle, with the different developmental stages in evolutionary history identified. The reptile brain is shown in dark grey, while the early mammalian brain is shown in light grey. The most developed mammalian brain – i.e. the human brain – is shown in white. Several brain structures that have central and definable roles are mentioned by name.

The mammalian brain

All mammalian brains consist of so-called grey and white matter. The grey matter – which isn't actually grey, but pink – is home to the neural cell bodies and the synapses, where signals are transferred between the neurons. The white matter, which serves as the highway for those signals, runs through long, wire-like axons. Like all other wiring, the brain's wires must be insulated to function properly. The insulating material is called myelin, which is white because of its high fat content. We find grey matter in the cerebral cortex – around the cerebrum and the cerebellum – but there are also islands of it in the middle of the brain.

The human limbic system consists of all of the structures that appeared in the first mammalian brains as well as the oldest parts of the cerebral cortex and islands of grey matter made up of neurons. Many of these neuron islands (also known as nuclei) are vital for a number of the body's basic functions – crucial evolutionary instincts that are sometimes referred to as the 'Four Fs': fighting, fleeing, feeding and fucking.

One particularly important limbic nucleus is the amygdala, which is located inside the temple (see Fig. 1). 'Amygdala' is the Greek word for almond – early anatomists named the structures in the brain according to what they resembled – and it plays a central role whenever the human body is engaged in one of the first two Fs. Its neurons not only cause you to blurt out a few choice words when you run for a bus and the driver pulls away just as you reach the stop, but also make you feel worked up all over again when you relate the story later the same day. The amygdala is also important for motivation, so it is at least partly to blame when you break into a flat-out sprint to try to catch that bus, even though you know the next one will be along

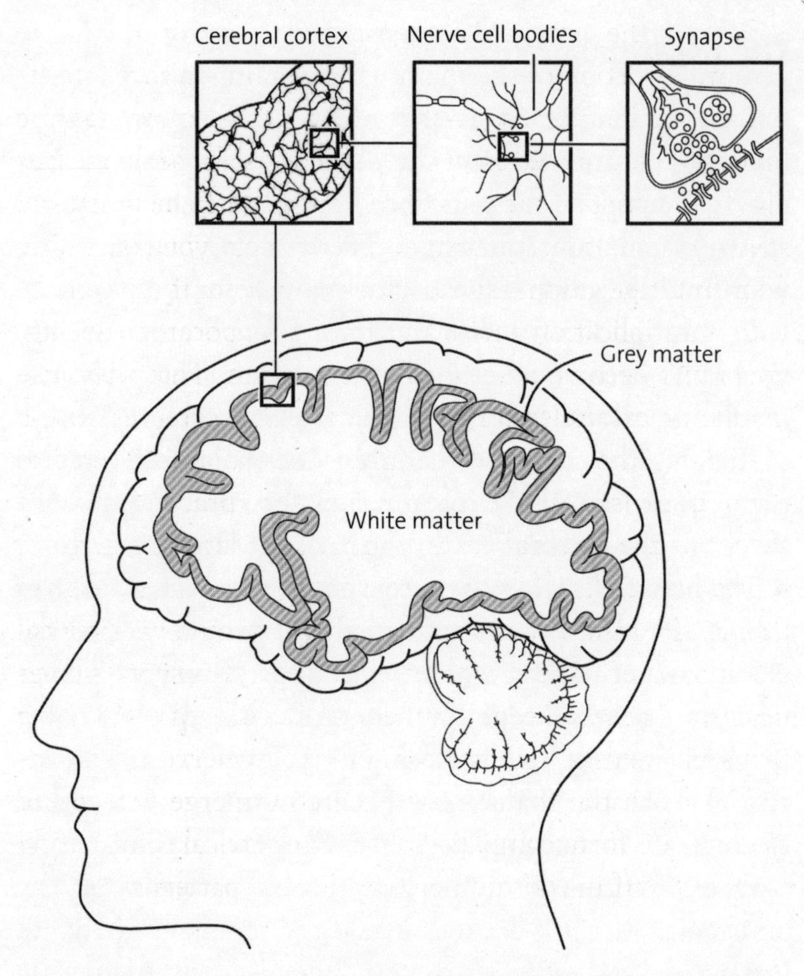

Figure 2: The cerebral cortex is made up of grey matter, and it is also where we find all the neurons and the contact points between them – the synapses. Inside the grey matter we find white matter, which is made up of insulated neuron axons.

in a few minutes. Moreover, when you're walking home in the dark and hear footsteps behind you and pick up your pace a bit, that's your amygdala at work again. Even if you were in a safe environment and didn't have anything to fear, you would feel absolute terror if your amygdala were electrically stimulated.

Behind the amygdala sits another part of the more primitive portion of the mammalian brain – a three–four-centimetre-long, sausage-shaped structure known as the hippocampus, named after the Greek word for seahorse (see Fig. 1). The hippocampus is important for both memory and spatial orientation: for instance, it can help you remember your multiplication tables. However, even if you recite your multiplication tables until your hippocampus aches, you won't become a better mathematician. That's because mathematical understanding resides in the cerebral cortex.

Right in the centre of the brain, the thalamus straddles the midline (see Fig. 1). It sends signals to practically every corner of the cerebral cortex with all the latest news from the body's senses. If we were to compare brain structures to people, then the left and right thalamus would be the local gossipmonger – the neighbour who knows everyone's business, the one who keeps a finger on the pulse of everything that's happening. Extensive highways of nerve cell axons run through the thalamus, where they merge with other roadways to form complex circuits of electrical information that zoom off in coordinated, rhythmical patterns.

Ingenious apes

A long time ago, our ancestors lived in the treetops of Africa's jungles, and they stayed there right until the moment when the climate changed. Back then, the earth's climate was like a roller coaster, with a series of alternating ice ages and heatwaves. While these fluctuations eventually forced our ancestors down from the trees, they were not extreme enough to wipe them out. The earth-bound apes quickly developed larger brains while also retaining both the reptile brain and the limbic system. The increase in volume was entirely due to enlargement of the cerebral cortex.

The early hominids who walked on two legs across the African savannah almost four million years ago had brains that weighed about four hundred grams. Although their hands could now do something other than cling to branches, they didn't hold tools until 'the handyman' – *Homo habilis* – appeared two million years ago. By then, the size of the hominid brain had increased to a good six hundred grams. Using tools was undoubtedly a breakthrough, even though *Homo habilis* was hardly sophisticated – he mostly just grabbed rocks and hit stuff with them. Moreover, these weren't the only creatures to make such a breakthrough: dolphins use bits of sea sponge to protect their beaks as they search for prey on the sea floor; cactus sparrows use thorns to flick insect larvae out of holes; and chimpanzees use branches to scoop termites out of tree trunks. Using tools to extract termites might be impressive, but it's quite a long way from writing a symphony. So, something else must have happened during the course of human evolutionary history – something that made our brains unique in the animal kingdom.

Another million years passed and *Homo habilis* gave way to *Homo erectus* – 'the upright man' – who was less governed by the primitive parts of his brain, which now weighed about a kilogram in total. Instead of running away from fire, he understood that he could use it for light, heat and protection on his journey further into the world. He also started to hunt. Finally, *Homo sapiens* – 'the thinking man' – evolved some 200,000 years ago, boasting a brain that weighs between 1200 and 1400 grams – three times heavier than those of the hominids who had first walked on two feet a mere 3.8 million years earlier.

In parallel with this steady increase in brain size, humans developed a level of intelligence that leaves other species far behind. However, this isn't just a matter of size. The brains of dolphins, chimpanzees and even cows are all similar in size to our own, but they cannot match our creativity or innovation.

Why isn't it enough to have the biggest brain?

Some animals have even bigger brains than we do. For instance, a blue whale's brain weighs a whopping eight kilograms . . . but it's housed in a body that weighs 170 tons! In general in the animal kingdom, the bigger the body, the bigger the brain. So what about our close relative the gorilla, which is two to three times the size of the average human? Are gorillas' brains correspondingly larger than ours? Actually, the opposite is true: our brain is two to three times larger than a gorilla's. Indeed, only whales and elephants have larger brains than humans: that is, the largest animals in water and on land, respectively. Relative to body size, then, the human brain is the largest of any animal.

Moreover, having a brain that weighs eight kilograms doesn't help the blue whale in terms of intelligence, since IQ isn't measured in kilograms. Two brains of identical size don't necessarily have the same number of neurons or an equal capacity for complex thought. A classic example is Albert Einstein, who, despite being the father of the theory of relativity and the winner of a Nobel Prize in physics, had a brain that was 20 per cent smaller than average. We know the precise size thanks to an unethical doctor. Einstein wanted to be cremated after his death and have his ashes spread somewhere that would make idol worship impossible. However, this wish was not fulfilled, because the doctor who conducted the autopsy stole his brain and took it home with him!

Different species' brains are also not built the same way. In primates (humans and apes), the size of the neurons themselves is the same whether the brain weighs eighty grams or a kilogram. Therefore, a one-kilogram brain would have ten times as many neurons as a one-hundred-gram brain, plain and simple. However, in rodents, larger brains have larger neurons. So, to get the same tenfold increase, a rodent's brain would need to be a whopping forty times heavier. Hence, a primate's brain will always have more neurons than an equal-sized rodent's brain. If a rat's brain had as many neurons as a human brain, it would weigh thirty-five kilograms! Thus, we don't just have the largest brain relative to body size; we also have far more neurons per gram than any other animal.

On the other hand, although rodents' brains and primates' brains differ significantly, the basic principles are identical. The neurons appear to talk to each other the same way. This is why rats and mice are often used in research into how the human brain works.

The unfinished brain

We couldn't really have much bigger brains because of the way we are constructed. There simply isn't any more room in our skull. Even though the cerebral cortex curls neatly to fit, the skull is already so large that human babies don't have much wiggle room at birth. If a baby doesn't turn the right way at the right time, there can be serious trouble. Nevertheless, the brain is still unfinished at that moment, to enable the head to make it through the birth canal. The disadvantage of this is that human children are entirely dependent on their parents for a very long time. Effectively, we give birth to small, helpless creatures whose brains continue to develop once they're outside the uterus, and as a result we need to put a lot of energy into raising every individual.

However, even though humans are vulnerable and need protection for at least the first decade of life, the human population has grown until there are now more than seven billion of us. In the last half century alone, the number of humans on earth has doubled. So, how have physically weak, naked apes who give birth to helpless infants achieved such a dominant position? We don't run particularly fast or swim particularly well, and our night vision is terrible. Therefore, we would seem to be at a significant disadvantage when compared to other animals – hunters and hunted alike. Successful predators tend to have powerful jaws, multiple rows of sharp teeth, paralysing venom or great speed, while prey animals protect themselves with thick armoured skin, camouflage or exceptional hearing. We have none of these assets.

Intelligence is an art

From an anatomical perspective, humans were entirely 'modern' 150,000 years ago, although there is no concrete evidence that these first *Homo sapiens* were capable of abstract or symbolic thought. However, we know that humans began producing works of art, jewellery and advanced tools, like canteens and fish hooks, about 40,000 years ago. We made these tools to compensate for our lack of physical attributes, but what led to this sudden burst of creativity? The only explanation is that there must have been a considerable change in the human brain around that time. Maybe this was due to genetic mutation? Or perhaps it was linked to Darwin's principle of 'survival of the fittest': perhaps the most creative and intelligent members of the species were considered the most attractive, and thereby had the greatest chance of passing on their genes? No one knows for sure.

There is a big difference between hitting stuff with a rock and building pyramids. Egypt's pyramids were erected some four thousand years ago, and the largest one consists of about 2.3 million stone blocks. Each of those blocks weighs about 2.5 tons, and they are so square that there is no more than a 0.1 per cent difference in the lengths of the opposite sides. It wasn't primarily muscle power that moved them into position, it was engineering: it was the brain. A couple of thousand years later, Eratosthenes calculated the earth's circumference so precisely that his conclusion deviates by only 2 per cent from the figure we use today; he did it by measuring the shadows the sun cast in Syene (modern-day Aswan) and Alexandria. Another couple of thousand years later, we are sending robots to Mars.

From the treetops to prime-time TV

It's not just the size of the human brain that's important, but also which parts of it account for that size. As we have seen, humans are more intelligent than other animals because of both our large brain size relative to our bodies and our unusually large cerebral cortex. On average, of the 86 billion neurons in the human brain, 16 billion are in the cerebral cortex. No other species has more neurons in its cerebral cortex than humans. It is the seat of our thoughts, language, personality and problem-solving ability. In short, it is what makes people people.

We outshine the other animals because of our cerebral cortex, even as we sit on our sofas watching TV. A satirical news anchor might report a story with a serious face and then introduce footage of the exact opposite of what he has just said. This causes us to start laughing, because our brains understand parody. Our highly developed cerebral cortex allows us not only to read the news anchor's emotions, but to interpret the hidden meaning behind what we're hearing and understand that something that was seemingly stated in all seriousness was actually sarcasm. Do you feel very superior as you sit there on your sofa, deciphering all of this in a matter of milliseconds? Well, you should, because only a creature with an exceptional brain can develop humour and language.

Other animals communicate too, of course, but their exchanges are limited to warnings about danger, along with expressions of pleasure, hunger and the desire for a mate. Because humans can read, write and talk, we have few limitations as to what we can communicate. We can use these sophisticated tools to write plays or compose operatic arias ... or to understand someone else's joke.

Not stronger but smarter

Everything has to do with sex, at least in evolutionary terms. Early humans developed their complex brains because this gave them an advantage in spreading their genes. Those who couldn't solve problems quickly or failed to learn from their mistakes didn't survive long enough to procreate. Nowadays, our brains help us to handle difficult situations and allow us to make friends, not enemies. They make it possible for us to save money over time, so that we can turn our attention to more ambitious goals. If you're clever and play your cards right, you can find a good partner, a rewarding job and supportive friends. In general, bright brains are more attractive than dull brains. Therefore, the net effect of our evolutionary journey has been the development of a species that is not stronger or faster, but smarter.

Chapter 2

Hunting for the Personality

Cogito ergo sum. The French philosopher René Descartes's familiar statement is usually translated as: 'I think, therefore I am.' But who are you? What makes you you? Personality is a combination of how you see yourself and how others see you. You aren't just what you think and feel, but also what you do and how you behave. But is this 'self' constant?

Philosophers aren't the only ones trying to find the answers to these questions any more. Brain researchers are searching for them, too. As with everything else in medicine, the question of nature versus nurture crops up here. And, as ever, we can say that both are crucial. Anyone with a sibling knows that two people who are raised in an identical environment do not share the same personality. Siblings who grow up together can develop completely different temperaments, values and attitudes. But the environment plays an important role, too. How a child is raised and how their role models behave both contribute to changes in that child's brain. Children see, hear and learn. Unfortunately, those who grow up in violent surroundings are likely to become violent themselves. Those with religious parents and friends are likely to become religious themselves. And those who are raised in a home filled with empathy and respect have a high chance of becoming empathetic themselves. However, these developments generally do not continue after childhood – most people's personalities are set by the time they become adults.

The seat of the soul

Descartes did more than point out that we are because we think. He was also convinced that the body and the soul are separate entities, with the latter non-physical. He argued that all of the sensory information we receive about the world is collected by the pineal gland – named because of its resemblance to a pine cone (see Fig. 3) – and then forwarded to the intangible soul.

But what is the soul? If it is 'I' – the sum total of what we think, feel, believe and do – then that isn't too far from what we now call the personality.

In 1848, two hundred years after Descartes proposed his theory, the railway worker Phineas Gage's tragic fate helped us say with certainty that the seat of the human soul is indeed in the brain. However, it is not in the pineal gland. In a freak work accident, an explosion forced an eleven-inch iron rod through Gage's left cheek, behind his left eye, through his left frontal lobe, and out of the top of his skull. The entrance and exit wounds were treated within an hour, and six months later Gage seemed to have made a remarkable physical recovery, save for loss of vision and drooping of the eyelid in his left eye.

Yet the accident had altered his personality completely. The iron rod had destroyed the front section of his frontal lobe, with the result that he was no longer able to keep appointments or control his temper. Unsurprisingly, he was

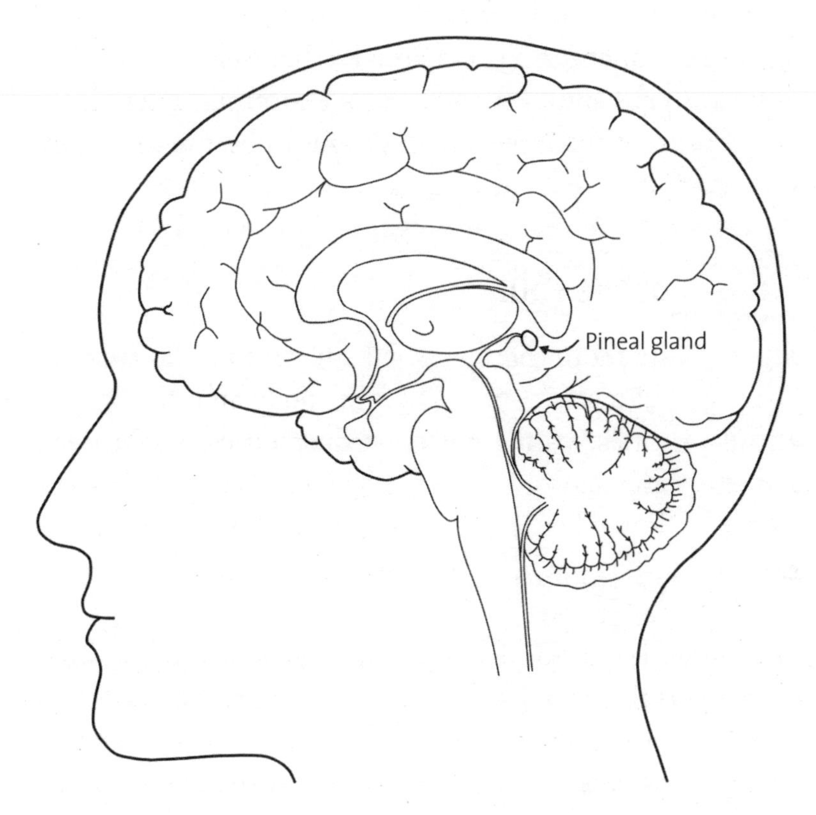

Figure 3: The right half of the brain, seen from the middle. The pineal gland – or epiphysis cerebri – is on the midline, towards the rear.

also unable to hold down a job. He died twelve years after the accident, steeped in alcohol, alone and abandoned by society.

Gage's story has become a classic case study for brain researchers because this was the first time in history that a link was established between a change in personality and a traumatic brain injury. Notwithstanding the prevailing view in the nineteenth century that the human personality was both unassailable and intangible, the consequences of a serious injury to the frontal lobe were suddenly undeniable. It was apparent that Descartes hadn't been right about everything. The self was physical.

Many other philosophers, theologians and scientists throughout history have claimed to know the location of the seat of the soul. For instance, as far back as the second century AD, the Greek physician Galen argued that the soul was contained in the liquid that surrounds the brain – the cerebrospinal fluid. It's a good sign that we can laugh at such suggestions now. Science has progressed and our knowledge is greater than it has ever been. For example, while it can no longer be called the seat of the soul, we know that the pineal gland has an important role to play because it produces the hormone that regulates our circadian rhythm.

A place for everything

The cerebral cortex is divided into lobes according to where the various parts are located within the cranium (see Fig. 4). Although many characteristics are linked to one specific area (or one lobe) of the brain, the various lobes do not operate in isolation. All of the brain's neurons have to be part of a network in order to function correctly. So, even characteristics that are considered to be located in particular centres depend on cooperation between groups of neurons in other parts of the brain.

The parietal lobe sits inside the parietal bone in the crown of the head and enables us to feel someone stroking our cheek or tears running down our face when we cry. The temporal lobe is inside the temple; it is important for memory, smell and hearing. The occipital lobe is crucial to our ability to see. The frontal lobe helps all mammals, including humans, to control their movements.

Every human has two speech areas in their dominant brain hemisphere. The left hemisphere is dominant in all

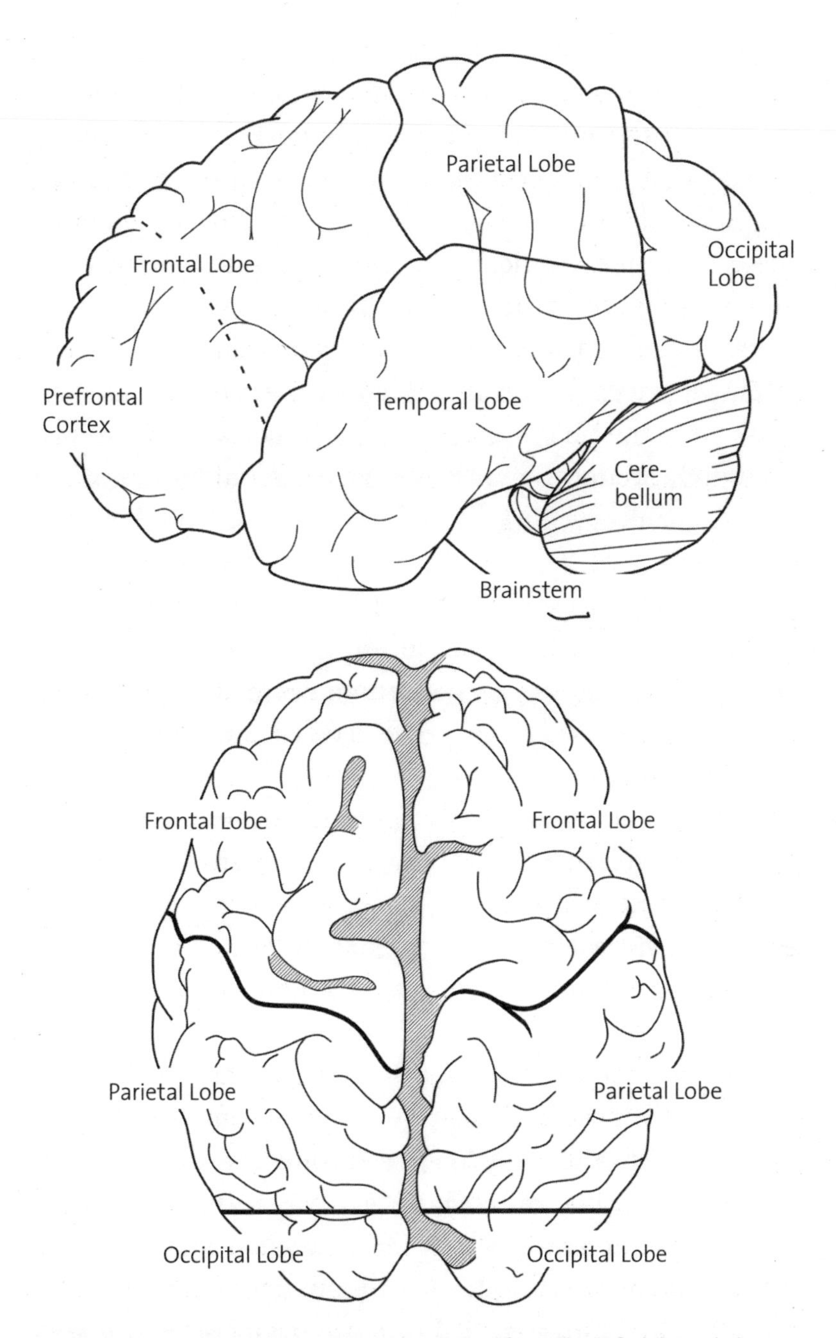

Figure 4: The lobes of the human brain shown from the left side and above. The second image shows that we have two of everything – for instance, a right frontal lobe and a left frontal lobe.

right-handed people, but the language areas are located there among 70 per cent of left-handers, too. The language area that enables us to generate language is located in the frontal lobe, while the one that allows us to understand language is located between the temporal and parietal lobes. If the latter language area is damaged, you may well talk incessantly but neither you nor those listening will be able to understand what you are saying because your brain will just invent words that don't exist. And you won't understand what other people are saying, either. On the other hand, if the language area in your frontal lobe is damaged, you will understand everything but will be unable to find the words to respond.

Other vital functions are also located in the frontal lobe, and specifically in the prefrontal cortex. This is the most recent addition to the human brain, and not just in an evolutionary sense; it is also the last part to develop as we grow up.

Altogether, the various regions of the cerebral cortex give us the ability to think analytically, to understand the consequences of our actions and to plan for the future. They also allow us to become mathematicians, poets and composers.

The frontal lobe

Damage to the frontal lobe results in the loss of many personality traits, which makes everyone who suffers such an injury similar to each other. But what does a healthy frontal lobe actually do for you? First, it allows you to plan for the future. Losing this ability can have devastating consequences, as it did for Phineas Gage. Why would anyone go to the trouble of arriving at work on time if he wasn't worried about losing his job tomorrow? So, your frontal lobe helps you carry out plans, but it also constrains you. In other words, without a functioning frontal lobe, you lose self-control and will probably end up doing something you regret. Or at least *would have* regretted, because damage to the frontal lobe also has a serious impact on self-awareness. Gage became unemotional, indifferent and apathetic because he no longer had any interest in or understanding of other people, so he hurt and injured those around him.

When brain researchers want to test a person to see if they have lost the ability to understand that the rules of a game have changed, many of them will pull out a deck of cards. Based on the tester's responses, the subject is encouraged to sort the cards in a particular way. Eventually, they work out that the black and red cards should go in separate piles. After a while, though, the tester no longer wants spades to be placed on top of clubs. It's normal for the subject to be confused at this point, but then they will figure out that the rules of the game have changed and they will start sorting the cards by suit. By contrast, someone with frontal lobe damage will be unable to accept that the rules have changed; instead, they will continue to group clubs with spades over and over again.

The director behind your forehead

The frontal lobe does more than organise personality traits, however. Without it, we wouldn't be able to wiggle a finger because all human movement is governed by the rear of the frontal lobe. Meanwhile, the foremost part of the lobe – the prefrontal cortex – controls our morality and humour. It helps us evaluate the possible consequences of every potential action, which allows us to align our behaviour with generally accepted norms and guidelines. It also contains the working memory, which helps us hold on to sensory data while we evaluate it and link it to ongoing processes and previously stored information.

So the prefrontal cortex is like a director controlling the brain. It is a command centre that collates all of the information relating to one's 'self' into an overall picture. It receives nerve signals from other parts of the cerebral cortex as well as from deep parts of the reptile brain. It also has a supervisory role by linking complex functions like memory, intellect and emotions. This provides the foundation for personality, conscience and other brain functions that are unique to *Homo sapiens* and distinguish us from other animals.

Personality in every corner of the brain

Although the frontal lobes are of paramount importance, a complex function like personality requires extensive cooperation between *all* of the brain's regions. Most people respond to the question 'Who are you?' by stating their name, age, where they live and their occupation. Your parietal lobes are in charge of this sort of factual information (see Fig. 5). They also make it possible for you to know that the hands which are holding this book are yours. So, if you had a stroke that affected one of your parietal lobes, you might look at your arm and think that it belonged to someone else! In other words, the parietal lobe helps you recognise yourself – and not just your physical self, but also how you think and view your inner self.

The centre for emotions and memory is located in the temporal lobe. If you pulled this lobe aside, you would see a region called the insula (see Fig. 5). While the parietal lobe helps you understand that parts of your body – such as your arms and legs – belong to you, the insula enables you to appreciate that your memories are your own and allows you to recognise yourself in photographs. You also use this area of the cortex when searching for words to characterise yourself.

Scientists used to think that the cerebellum (part of the reptile brain, as we saw in Chapter 1) only coordinated movement, but recent research suggests that it also plays an important role in regulating certain personality traits. For instance, without a functioning cerebellum, you would always do and say the first thing that popped into your head. You would have no safety catch to prevent you from putting your foot in your mouth – mirroring the problems that arise with a frontal lobe injury. You would also become

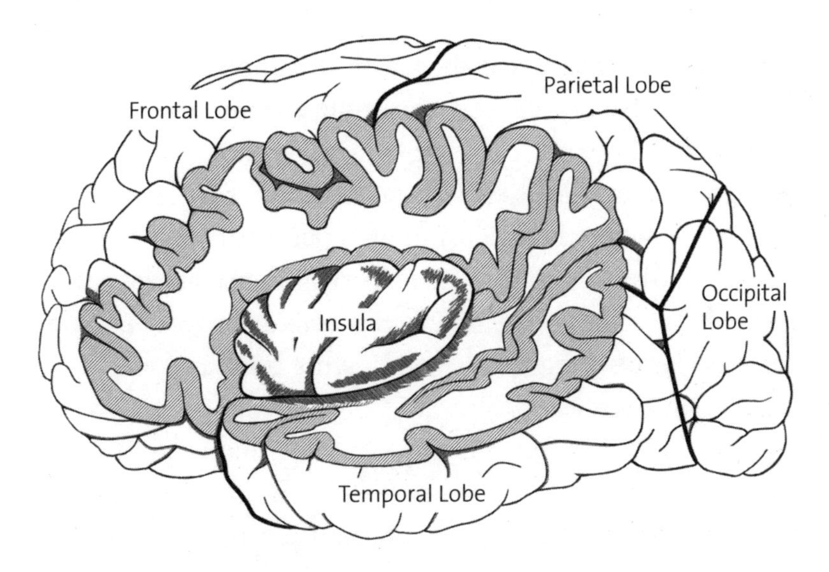

Figure 5: The left half of the brain, seen from the side, with the various lobes labelled. Parts of the brain have been omitted to reveal the insula – the area of cortex behind the temporal lobe.

emotionally unstable as your mood fluctuated between joy, grief and aggression without warning.

Personality is also related to our attitudes and choices. Researchers have found that we make a decision almost a full second before we are aware of reaching that decision. This doesn't mean that someone else makes the decision for you; it is simply that your consciousness is not involved at the start of the process. We like to think that we consciously choose to lift an arm and then lift it, but in reality the movement is planned before we know that the decision has been made.

Most of the studies on conscious choice have been quite simplistic. For instance, a research subject might be told to push a button with either her left or right hand while she watches a clock. The subject is supposed to notice where the second hand is at the precise moment when she makes the decision to push the button, but before she starts moving her arm. If electrodes have been applied to the subject's head, it is possible to watch her brain function and predict which hand she will choose before she believes she has made the decision.

As yet, no studies have attempted to explore more complicated choices, such as opting for a particular career or life partner. However, even if it turns out that some aspects of these choices are made before you become aware of them, the final decision is still *yours*. You are your brain.

Split brain, split personality?

What are the consequences of the fact that the human personality isn't one thing located in one place? As we saw above, almost all of our brain's lobes make some sort of contribution to the various traits that comprise the personality, and they are in constant communication with one another. The information that passes between the right and left hemispheres travels through the corpus callosum – a several-hundred-million-lane, white-matter superhighway that sits directly on the midline (see Fig. 6).

When treating a patient with severe epilepsy, doctors may reluctantly choose to cut the corpus callosum in order to prevent the condition from afflicting both hemispheres. After this operation, some patients are left feeling that they have two brains that think differently from each other. For instance, one half may want to remove the patient's trousers, while the other half will want to keep them on. The end result is that one arm will try to pull the trousers up

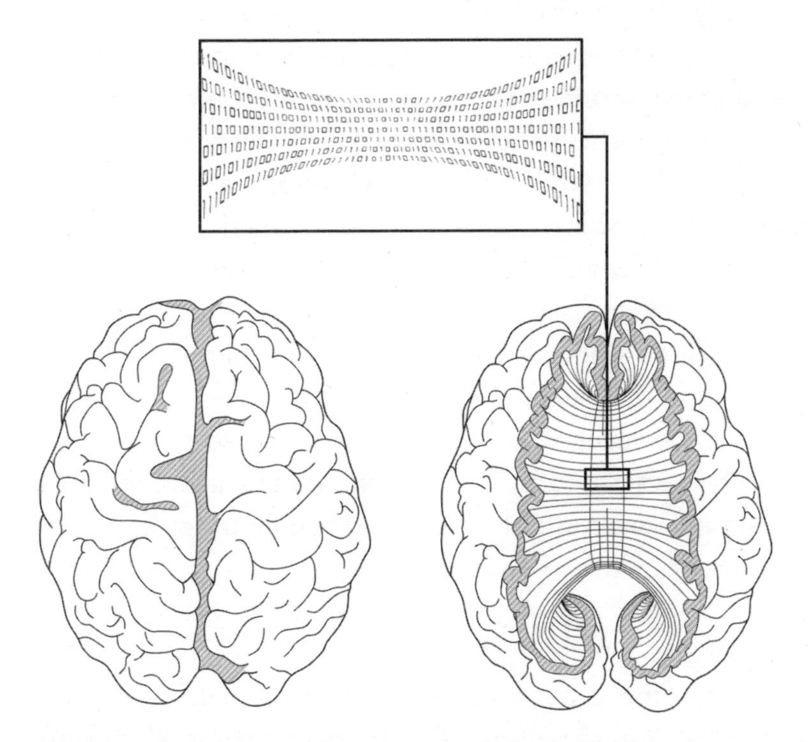

Figure 6: The brain on the right has been cut open to reveal the corpus callosum – the communication route between the right and left hemispheres.

while the other tries to pull them down. And it's not just movement that's affected: each hemisphere has its own thoughts, feelings, experiences and memories. In short, one person has two different minds.

The question is: does a person with a split brain have a genuinely split personality? Actually, decades of research suggest that, while people with split brains often have two distinct personalities, these are at least quite similar to each other. Maybe that's not so strange, given that surgeons usually don't perform the operation until the patient is an adult.

Dr Jekyll and Mr Hyde

The fancy, scientific term 'dissociation' describes something that is completely different from a split brain. You've probably experienced a mild form of this yourself: you don't hear what someone is saying because you're focusing on something else. Hopefully, you haven't experienced the more severe form. This involves two (or more) distinct consciousnesses that are never present at the same time – a number of personalities with their own preferences and behavioural patterns in a single body. Each personality has its own memory and cannot remember anything that happens to the other personalities. In both popular literature and scientific studies, the personalities tend to be at least different from each other and often polar opposites.

In many ways, then, the story of Dr Jekyll and Mr Hyde is a realistic depiction of a split personality: Jekyll is a pleasant, caring and popular doctor, while Hyde is the complete opposite. However, in Robert Louis Stevenson's novella, the split personality occurs as a result of one of Jekyll's own experiments, which is obviously not the case for real-life patients with dissociative identity disorder.

Multitasking

We live in a society that's obsessed with efficiency. It's not enough to complete one task, then move on to the next. We are placed in open-plan offices where we are expected to respond to emails, answer the phone and write a report all at the same time. Multitasking is the modern way. Multitasking is the future. Or is it?

In reality, no one can do two things simultaneously, because the brain is able to focus on only one task at a time. When you think you're reading a report while also ordering a takeaway, your brain is actually switching back and forth at great speed between reading and barking your order down the phone. This is an impressive feat and you don't even notice that your brain is doing it. But the end result is that it takes longer to complete the two tasks than if you'd ordered the food and *then* read the report. Thus, if you want to be truly efficient, you should learn to prioritise, not try to multitask.

Indeed, if you try to do something while still working on something else, you run the risk of temporary brain paralysis, because the prefrontal cortex can't just spontaneously shift its focus – there's always a pause as you switch between one and the other. This is especially true if the two tasks are similar, because they will compete for the same neural networks. For instance, trying to listen to the news on the radio and read a book at the same time requires activity in overlapping regions of the brain, which makes it much harder to do than listening to an audio guide in a gallery while looking at the exhibits.

Remember, though, trying to do two tasks at once *always* has an impact on the amount of attention we are able to devote to each of them, even if they are very different. It has been estimated that a driver who is talking on their mobile phone drives as inattentively as someone with a blood alcohol content of 0.08 per cent – the drink-drive limit in the UK. And that applies whether they're using a hands-free device or not.

So, increase your efficiency by focusing on one thing at a time. Finish reading your emails, *then* pick up the phone to make that important call.

You can change yourself . . . but only a little

Synapses (the connections between the neurons), membrane potential (essentially the difference in voltage between the inside and outside of a cell) and neurotransmitters (chemical messengers) all contribute to the human personality. Our thoughts, feelings and volition arise from the chemical and physical processes that take place within, among and between them. So we are biology, but we are not slaves to it. The brain can be influenced, attitudes can be changed, bad habits can be broken and temper can be controlled. If one part of your brain sends a signal to the language centre that the kettle you have just touched is so hot that you should blurt out some choice words, your frontal lobe can step in to stop you swearing in front of your children. Or, if you decide to rewrite an angry email from scratch rather than send it to your boss and to hell with the consequences, you can thank the part of the cortex that sits just behind your forehead.

If you have a spouse who believes that dirty clothes should be left in the middle of the bedroom floor rather than put in the laundry basket (as I do!), remember that everyone's brain is plastic, so habits can be altered throughout life. However, if you hope to encourage radical change in your partner, it might be time to start looking for someone else. The brain we have at birth and our upbringing play such fundamental roles in defining the human personality that it remains surprisingly constant, unless the brain is physically damaged in some way. A few traits may be modified, but in general you shouldn't expect to achieve much more than that.

Herd-wired brains

People are pack animals. We all have brains that allow us to cooperate and obey orders. This is one reason why we are able to live in relatively peaceful societies. But what would happen if we found ourselves in an environment where the norms were destructive and the leader was even worse? What would we do if the traits that have helped us get along with each other for thousands of years were suddenly exploited and used to our own detriment?

A young boy grew up in the 1930s in a poor household in Indiana with a hard-working mother and an alcoholic father. The neighbourhood kids kept their distance as he seemed to be obsessed with religion and death: he stabbed a cat to death in the street just because he wanted to perform a funeral service. Unsurprisingly, the boy started to feel increasingly excluded and ostracised. During his teenage years, even though he was white, he identified with the similarly excluded African-American community. In his early twenties, he started his own religious community, which he called the Peoples Temple. His name was Jim Jones.

The new church was open to all, regardless of race or background, and the poor, isolated boy became a charismatic religious leader to several thousand devotees. Over time, the congregation began to function increasingly like a cult, with the members living, sleeping and working on site and shunning all contact with outsiders. Jones became an autocrat who ordered marriages between members of the group and cracked down hard on any criticism. As the US authorities grew increasingly concerned about the church's activities, he decided to move the whole congregation to a newly constructed compound in Guyana that he named Jonestown. Just over a year later, on 18 November 1978, he instructed all 909 cult members to commit suicide. Parents forced their children to drink poison before taking it themselves. It was the largest mass suicide in history.

Why didn't the members of the congregation rebel against their obviously deranged leader? What had happened to the attitudes and values that had once defined their personalities as individuals? A couple of decades earlier, at the conclusion of the Korean War, the term 'brainwashed' was coined to describe US prisoners of war who had become ardent supporters of communism during their captivity. Unfortunately, since then, there has been relatively little research into this phenomenon. However, a number of studies have explored group thinking.

Time and again, this research has shown that ordinary, decent people can be persuaded to act in ways that they would never normally countenance. For instance, an American history teacher was finding it hard to explain how Hitler managed to gain the support of almost the whole German population in the 1930s, so he decided to demonstrate the process in action. He placed a small number of students in a group he called the 'Third Wave' that was based on two of the founding principles of the Nazi Party: strict discipline and strong community. Over the course of the next five days, the group recruited ever more members and started to spiral out of control, so the teacher decided to end the experiment. He summoned all of the members – who already numbered in the hundreds – to what was billed as a presentation by their supreme leader. When the students arrived, the teacher showed them a picture of Hitler. Many of the members started to cry when they realised the sort of organisation they had been so eager to join.

When I saw the movie *The Third Wave*, which is based on this experiment, I was sure I would have been one of the few to resist the pressure to join the group. I was convinced that my brain would have had the strength to stand alone. But would I really have been as resolute as I'd like to think?

In another experiment, US social psychology researcher Stanley Milgram found that 65 per cent of normal, upstanding people were willing to hurt a fellow human being if ordered to do so by an authority figure. In this first part of the study, the subjects were assured that the person issuing the order bore full responsibility for the consequences of their actions, and they did not have to witness those consequences as they were tested in a separate room from the 'victim'. The figure rose to 90 per cent if the subjects did not have to inflict any pain themselves, but rather instructed someone else to do the dirty work for them. In the second part of the study, the subject was placed in the same room as a person who was pleading for mercy and supposedly receiving a painful electric shock every time a button was pressed (actually an actor who felt no pain at all). Rather more of the subjects refused to push the button in this instance.

Milgram's study revealed no gender differences. All of the subjects, including those who always obeyed, emphasised that they disliked the situation in which they had been placed. Nevertheless, although their brains produced stress hormones that caused them to sweat and stutter, the majority still did whatever they were instructed to do.

Many people believe that NASA's two most serious accidents – the *Challenger* and *Columbia* space shuttle tragedies – could have been avoided if someone had dared to stand up to the group. In both cases, questions had been raised about possible flaws in the design of the space shuttle, but the large majority's strong desire to avoid further delays caused those with serious doubts to remain silent.

Clearly, then, suppressing prior knowledge, following the rules and meeting the group's expectations is not always the best course of action. Although the urge to comply – which originates in the frontal lobes – is strong, it can be resisted, especially if you recognise the warning signs. Brain researcher Irving Janis advocates extra vigilance if you are working in a close-knit group that means a lot to you. In such situations, you might be particularly disinclined to offer opinions and information that could generate conflict with the other members of the group, and this is exacerbated if you are working under pressure, insulated from external opinions and have a strong leader. Nevertheless, warning lights should flash if people who express doubts about a project are instructed not to rock the boat. You should fight against this as self-censorship is usually unhealthy. If you and other group members share similar concerns but don't protest for fear of exclusion, the end result will be an illusion of unanimity in which possibly beneficial counter-arguments are never heard.

Bear this in mind the next time you wonder if you should bite your tongue or blurt out your opinions. Did I say 'in mind'? Of course, I mean in your frontal lobes.

Can your personality get sick?

Abnormal personality traits do not necessarily indicate that the people who have them are sick. They might simply *be* that way. Outside the broad spectrum of what is considered 'normal' lie the extremes of what we term 'personality disorders'. If you are wicked enough, egocentric enough, impulsive enough, dramatic enough or compulsive enough, then you are said to have a personality disorder. The entire nation of Norway became familiar with this concept during the trial of the terrorist and mass murderer Anders Behring Breivik. According to the first report on his mental health, he suffered from a *psychiatric* disorder and thus could not be held legally accountable for his actions. However, a second report stated that he suffered from *personality* disorders and therefore could stand trial.

A personality disorder is not a disease in the strict definition of the term; rather, it causes pronounced deviant personality traits that often create serious problems for the sufferer and those around them. Since the personality is shaped by the environment and fully realised only during adulthood, personality disorders are rarely diagnosed in children.

In a few cases, psychologists and psychiatrists have tried to exploit the brain's malleability in order to correct deviant personality traits. However, this requires a willingness to change on the part of the patient. Unfortunately, if you have one of the more common disorders – narcissistic personality disorder, which manifests as extreme self-absorption – you will be unable to acknowledge that you have any sort of problem, let alone that you would benefit from treatment. An even more well-known disorder used to be called psychopathy and is now termed antisocial personality disorder.

The second report about Breivik's mental health claimed that he suffered from both of these disorders, which are characterised by a lack of empathy.

The psychological is physical

Psychologists and psychiatrists work with the psyche, while neurologists and neurosurgeons work with organic brain diseases – in other words, the physical. But why is the distinction so sharp when we know that the psychological is also physical and all of these professionals work with the brain?

The human personality is not supernatural; rather, it is an amalgamation of an individual's unique heredity and unique experiences, which have led to unique connections among the neurons in that individual's brain. Diseases that affect our feelings or alter our personality traits are generally known as mental illnesses. This maintains the sharp division Descartes drew between the physical and the mental. On the other hand, we are finding ever more physical causes for so-called mental illnesses. For instance, about half of the patients with frontal lobe dementia develop antisocial behaviour which is reminiscent of illnesses we have traditionally categorised as mental. They might start shoplifting or drink-driving and seem to develop a total disregard for the normal rules of society. As the disease progresses, the loss of neurons in their frontal and temporal lobes can be so pronounced that it eventually becomes visible to the naked eye. This obvious atrophying of the brain tissue makes it easy for us to understand the biological processes that are taking place. Similarly, we know that a tumour in the occipital lobe can cause blindness, while one in the frontal lobe can have a profound effect on the patient's personality.

Unfortunately, brain researchers have made far less

progress in their studies of classic mental illnesses, such as depression, anxiety and schizophrenia. One reason for this is that the diagnoses for all of these illnesses are based on symptoms alone. So, if a patient feels dejected for a long period of time, their doctor may well diagnose 'depression' and prescribe antidepressants. However, that catch-all term doesn't tell the full story. The patient's sense of dejection and despair is probably due to twenty, thirty or even thousands of disorders in their brain's chemistry.

The same is true for schizophrenia. Researchers are currently searching for what distinguishes people who hallucinate from those who don't, but they are hampered by the fact that the former group is far from uniform. Some hallucinators seem to have been born with a genetic disposition to the condition, whereas others apparently developed disorders in their brain chemistry due to drug abuse. Given such wide discrepancies in the study group, one can't expect a simple answer to the question: why does someone develop schizophrenia? Therefore, this field of study would benefit from more cooperation among psychologists, neurologists and brain researchers, rather than each profession championing their own explanatory model while denigrating the others.

Many mistakes have been made in our attempts to treat the most serious mental illnesses. The widespread use of lobotomisation in the 1940s and 1950s is one of the worst examples. 'Lobotomy' is used as an umbrella term to describe any one of a number of operations that aim to destroy the connections between the nerve cells in the prefrontal cortex. Aggressive patients would generally calm down dramatically after undergoing the procedure, but their personality changed in other ways, too. Typically, they would be left with a blunted emotional life and little in the way of self-control, spontaneity and/or self-knowledge. The fact that the Portuguese neurologist who originated the procedure won 1949's Nobel Prize in medicine for 'curing' hallucinating patients reveals how little we knew about the human mind less than seventy years ago.

Indeed, the prefrontal cortex remained an enigma until the late twentieth century, partly because it was considered relatively insignificant. It is unfortunate that the scientific community did not identify the frontal lobe as the seat of the personality immediately after Phineas Gage's accident in 1848. Medical history might have looked very different if it had, and we may have been much closer to solving the mystery of so-called mental illness.

Do animals have personalities?

We are much more sophisticated than any other animal. Nevertheless, all mammals have frontal lobes and thus certain personality traits. In humans, the frontal lobe accounts for 30 per cent of the total volume of the brain. It gives us our humour, self-image, morality and judgement, among many other attributes. In dogs, the frontal lobe accounts for only 5–6 per cent of the total volume, but that's sufficient for them to maintain focused attention.

Our memory allows us to keep track of the past and predict the future, which gives us the sense of being the same person over the course of a lifetime. We are constantly aware of ourselves. In the brainstem, the reticular activating system is a group of neurons that controls our attentiveness. It keeps us awake and activates our frontal lobes, so it is a prerequisite for human consciousness. The actual content of that consciousness, on the other hand, is controlled by the frontal lobes themselves. In other animals, memory and consciousness are not so closely linked. When it comes to an awareness of time, for example, they are entirely focused on the here and now. As far as we know, we are the only creatures who have a clear understanding of our own history.

Our sense of self grew out of the complex social lives of our ancestors. They formed small groups and shared whatever food they found, which demanded considerable self-control and cooperation. This was only possible due to their consciousness of the self. Some animals have developed something similar: chimpanzees, for example, are able to recognise themselves in a mirror. But their sense of self is much less developed than our own.

Personality tests

Many companies now use personality tests when assembling project teams with the aim of increasing cooperation within the group. The most commonly used is called the 'Big Five'. This five-factor model grades the candidates' answers and rates them on a scale for five different traits:

- energy and enjoyment of the company of others (extraversion);
- quality of social contact (agreeableness);
- self-discipline and need for order (conscientiousness);
- vulnerability and temperament (neuroticism); and
- values, reflection and information handling (openness).

Where each candidate falls on the scales for these factors – and various combinations of them – supposedly says something about their personality.

This test is widely used in both scientific research and the professional world, and the internet has any number of short versions that you might want to try. However, the results should always be interpreted with caution. Our behaviour does not follow one consistent pattern in every situation. People who have taken tests like the Big Five at work know that the results would have been completely different had they been asked about a family situation instead. If someone makes what may be interpreted as a sarcastic comment, should you ignore it, laugh it off or snap back? Your brain bombards you with all of these alternatives and more. Your frontal lobe will help you choose the best option, but its decision always depends on your current situation. For instance, you might not want to take the leadership role in some situations, yet you'll feel in total command in others.

The human personality is complex because the human brain is complex. We all have personality traits we can strengthen or suppress as required. Becoming more familiar with how your brain forms your personality will enhance your ability to control your negative impulses and help you understand how you relate to those around you. In the meantime, researchers around the world are working on increasing our knowledge of this enigmatic field.

Chapter 3

Memory and Learning

Learning and memory are the foundation stones of culture. Without learning, there would be no development. Without memory, we wouldn't recognise our family and friends.

People who have suffered damage to various parts of the brain, and consequent memory loss, have taught us much of what we know about human memory. The most famous patient, Henry Molaison – better known simply as H. M. – became a bona fide celebrity in the field of neuroscience. In 1933, at the age of seven, he suffered a head injury as a result of a bicycle accident, which led to epilepsy. This condition causes the sufferer to experience debilitating episodes because their neural activity is disrupted either in a specific lobe or throughout the whole brain. H. M.'s case was particularly severe: he would frequently lose consciousness and fall to the floor without warning prior to experiencing prolonged convulsions. Moreover, after every episode, he would suffer extreme tiredness and lethargy. Unsurprisingly, it was not long before he was unable to keep up in school.

Twenty years after the accident, in a desperate search for some sort of solution, H. M.'s family contacted one of the pre-eminent neurosurgeons of the day, who suggested that the trigger for H. M.'s unpredictable electrical impulses was probably located somewhere within his temporal lobes. Hence, with the family's permission, the surgeon removed sections from these lobes, and H. M.'s epilepsy improved significantly. However, he lost the ability to form new memories and was unable to use his memory to travel mentally through time and space. In other words, he was trapped in the here and now. He would greet everyone he met politely, and was more than happy to walk and talk with them. Yet, if the same person returned an hour later, H. M. would invariably reintroduce himself as he had no recollection of their previous meeting. This meant he was unusually patient with the researchers who subjected him to numerous tests over the course of the next fifty years because he undertook each test as if for the first time.

Short-term memory

In the Pixar film *Finding Nemo*, the title character's father joins up with Dory, a scatterbrained blue reef fish, to search for his son. Like H. M., Dory has trouble storing new memories. However, she does have some power of recollection because she remembers where she is when she reads the word 'Sydney' on a drain. She also constantly calls Nemo by similar-sounding names. H. M. wouldn't have been able to do that as he had absolutely no memory of any of the people he had met and should have been able to name. Yet both Dory and H. M. were able to form and express rational, coherent thoughts.

Before researchers began to study H. M., they thought that memory was a single entity. However, their observations indicated that a person can lose one part of the memory yet still retain another. Using this as their starting point, they gradually began to divide memory into short-term and long-term. H. M.'s short-term memory was left largely unaffected by the operation.

Many people use the term 'working memory' as a synonym for all short-term memory. Others think of it specifically as those parts of short-term memory that demand all of our focus, while the other parts are more passive and relate only to memory storage, which does not demand any conscious attention. Yet the difference between these two parts is so blurred that I and many others prefer to treat them as a whole.

The difference between short-term memory and long-term memory is not crystal clear either, but at least in this case a distinct anatomical difference was apparent after parts of H. M.'s temporal lobes were surgically removed. Following the operation, he could still remember random words and numbers for several minutes, as long as he wasn't distracted. Therefore, the researchers were able to say conclusively that short-term memory is not located in the temporal lobes. (Later research would reveal that it is actually located in the frontal lobe.) It's important for reasoning, making plans and formulating alternative solutions to problems, but H. M. showed that it's difficult to function properly with nothing but short-term memory.

Have you ever sat talking with a friend, only to find yourself distracted by a more interesting conversation on the neighbouring table? Nevertheless, you keep nodding and smiling until you hear your friend's intonation rise at the end of a sentence, at which point you suddenly realise that you have no idea what they have just asked you. Your working memory is limited. In order to remember something, first we have to process the information our senses provide by asking a series of questions. What's important to me?

What's missing? What would I like to know? Do I agree with the assumptions? Then, in order to recall that information later, we have to memorise it. Even though you heard the words as your friend spoke them, you weren't fully focused on what they were saying, so you neither processed nor memorised the sensory information you received. As your friend waits for an answer, your only option is to admit that you got distracted and ask them to repeat the question.

When my family gets together in the mountains during the Easter holidays, we play a game that involves memorising a number of objects or words in the space of a minute. We are a motley group of people between the ages of twenty and sixty from different backgrounds and with various levels of education. Yet there is strikingly little variation when it comes to remembering those objects and words in a given amount of time. Usually, most of us manage seven – the number of heavens in Allah's universe and the number of colours in a rainbow, but also, it turns out, the number things that an average human is able to process and remember at one time.

Long-term memory

Some people can remember much longer lists of words, however. By using brain scans, we can observe the activity in the innermost parts of these people's temporal lobes. When asked to memorise a long list of words, it appears that the words they heard first have already been stored in their long-term memory while more recent words are still in their working memory. The transition between the two seems to be quite fluid.

H. M. played another important role in increasing our understanding of human memory. In the early 1960s, he was asked to draw a star while looking at a mirror image of one, but without being able to see the page on which he was drawing. He did the best he could, but the results were terrible. The next day, the researchers asked him to try again. Predictably, H. M. said that he'd never done anything like this before, so the scientists repeated the same detailed instructions. Again, H. M. tried his best but struggled. Yet the end result was better, and his drawings continued to improve over subsequent days. So, even though H. M. couldn't remember performing the task, and had to be given the same instructions every day, it was as if his hand *did* remember.

In light of this experiment, long-term memory was divided into data memory and motor memory. When you try to ride a bike or swim for the first time, studying these skills or listening to someone explain the theory behind them isn't much use. The only thing that really helps is practice, practice and more practice. And the information you accumulate from all that practice is stored in your motor memory, otherwise known as implicit memory.

Meanwhile, data memory – also known as declarative or explicit memory – involves all of the factual information and experiences that you store away as memories. For instance, when you study multiplication tables or memorise a list of kings and queens, all of that information becomes part of your data memory, along with everything you've experienced in your life.

The hippocampus and its pals

The specific part of H. M.'s brain that was surgically removed in order to treat his epilepsy was the sausage-shaped structure known as the hippocampus (see Fig. 7).

Since the 1950s, we have known that memories are spread right across the cerebral cortex, and this was confirmed by the fact that H. M. could recall everything he had experienced until a couple of years before the operation: that is, he retained all of the memories he'd stored up to the age of twenty-five. Thereafter, however, he was unable to store any new ones and went through the rest of his life believing that he was still in his twenties. When he saw a picture of himself in later life, he thought it was a photograph of his father, even though he knew his father didn't wear glasses. He was also surprised by his own reflection each morning. All of this led the scientists who studied his case to reach the conclusion that the hippocampus must play an important role in the memory storage process.

We now know that if you are to remember what you experience, read or discuss, the hippocampus has to encode that information. Otherwise, it will simply disappear. The hippocampus receives signals from the olfactory (smell-related) cortex, the auditory (hearing-related) cortex, the visual (sight-related) cortex and the somatosensory (touch-related) cortex, as well as from the limbic system, where our emotions are generated. Once it has received this wealth of data, it creates a memory – or, more precisely, fragments of memory that can be amalgamated later.

The frontal lobe is the hippocampus's best friend, its wingman, the pal who tells it how to deal with the information it receives and what to forget. That's because the working memory in the frontal lobe processes all of the sensory information before sending it to the hippocampus for storage. Sometimes, though, the frontal lobe forgets what it should be doing and instead starts to babble with the hippocampus about holiday plans, dreams and all sorts of other things. When that happens, the hippocampus is unable to fulfil its

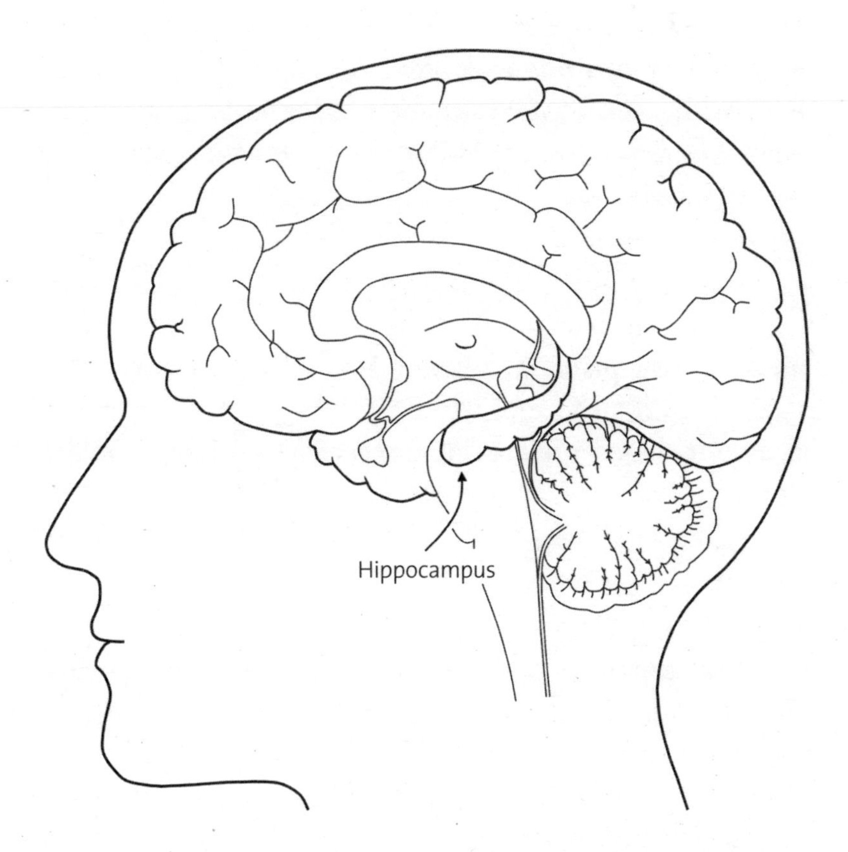

Figure 7: The right hemisphere of the brain, seen from the centre, with most of the left hemisphere's temporal lobe removed to reveal its hippocampus.

storage responsibilities. That's when you have to take control: for instance, you may have to read a chapter over and over again in order to force the frontal lobe to send the hippocampus all of the information it needs to create a new memory.

The cerebellum and the basal nuclei are other friends, although the hippocampus doesn't hang out with them very much (see Figs. 8 and 9). They are also involved with memory, but specifically with motor memory, rather than data memory. They work in unison to make us better pianists, better artists or better football players – provided we put in the necessary effort. So, if either one starts to malfunction, practice will no longer lead to mastery. In short, while the frontal lobe and the hippocampus help us remember *what*, the cerebellum and the basal nuclei help us remember *how*.

Remembering for the future

The memory's primary function is to improve our chances of survival. It's the tool we use to change and adapt our behaviour, based on previous experience. What should I

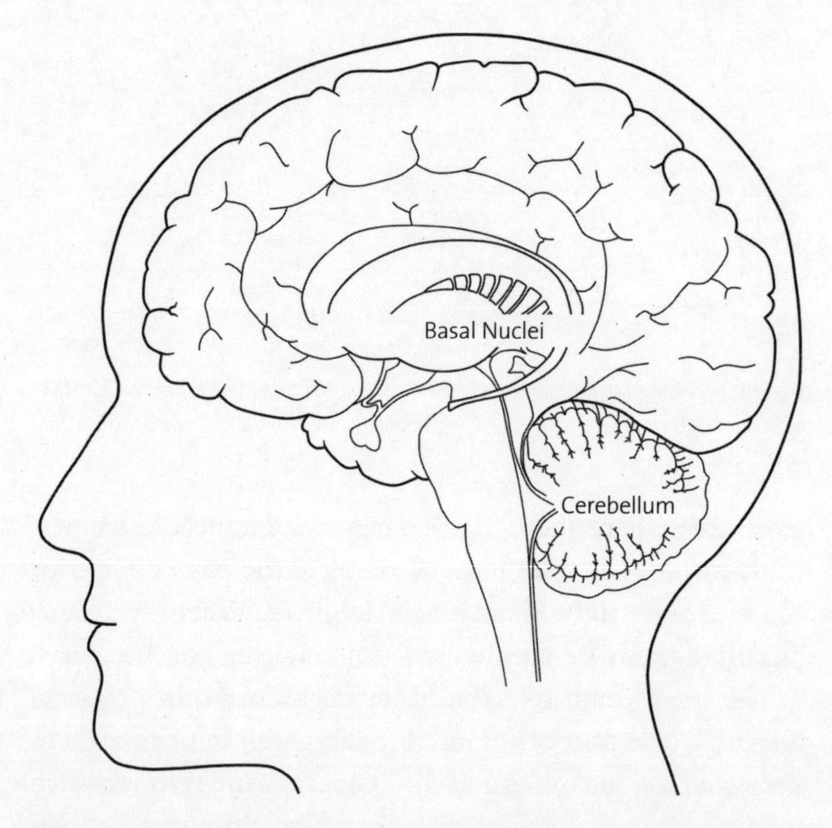

Figure 8: The brain's right hemisphere, seen from the centre, with most of the left hemisphere removed to reveal its basal nuclei – groups of nerve cells situated deep inside each half of the brain.

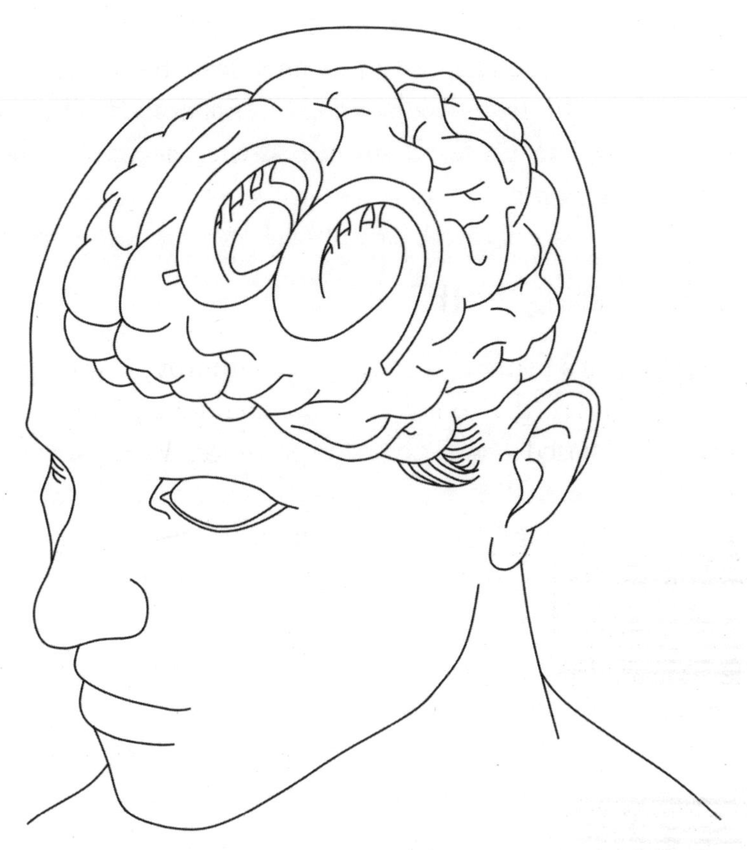

Figure 9: The brain seen from above and to the left, with each side's basal nuclei visible.

do now? Where should I go? What can I expect to happen? Our memory is not meant to recreate the past, but to help us make the right choices in the future. When we picture future actions or plan what we should do, we are able to form these scenarios in our mind because of our memories. Of course, memory does not provide a perfect, unchangeable image of the future; it can be constructed and reconstructed, depending on our knowledge of our surroundings.

An important part of this process takes place in the hippocampus, where expansive, interconnected scenes are assembled on the basis of what we have previously experienced. Therefore, in addition to losing their ability to store memories of the past, people who suffer damage to the hippocampus are unable to picture the future. Like H. M., they become locked into the here and now. They cannot travel mentally through time. A pair of healthy hippocampi allow the rest of us to undertake that kind of time travel.

Learning

Learning involves the acquisition of knowledge, while memory is related to storing it. In other words, without learning, you would have nothing to remember. On the other hand, memory is essential for all learning because you have to be able to store – and then retrieve – the information that you learn.

Many areas of the brain contribute to learning. For instance, the prefrontal cortex and the hypothalamus (our hormone centre) play important roles through a complex system of reward and punishment. In turn, sustained practice can have a profound – sometimes even a visible – effect on the areas of the cerebral cortex that are responsible for movement, as these develop in response to the challenges we set for ourselves. During surgery, it's important to have equally good use of both hands. I'm right-handed, so one of my colleagues suggested that I should try brushing my teeth with my left hand in order to train my brain to make better use of that hand. It was sound advice. Studies have shown that the part of the cortex that governs the left hand is larger in musicians who play the strings of their instruments with that hand when compared to individuals in a control group. And the difference is greatest in musicians who started playing at a young age.

Clowns and drooling dogs

While the Russian physician Ivan Pavlov was studying dogs' digestive systems and specifically what their saliva contained at various stages of a meal, he noticed that the dogs started drooling before he presented them with any food. That is, they started salivating when they thought they were about to be fed, even if they had nothing to eat. The mere sound of approaching footsteps was enough to set them off. This discovery prompted Pavlov to investigate the link between the stimulus and the physical response. He found that he could teach the dogs to associate almost anything with the prospect of receiving food. First, he made a specific sound, then gave the dogs food. After a while, the dogs started drooling whenever they heard the sound. This type of learning is called 'classical conditioning'.

When my younger sister was a child, she loved strawberry ice cream. All of the other children would choose chocolate, but she always had to have strawberry. One day, our grandparents allowed her to help herself to their strawberry ice cream. Like most children, she didn't know her own limits: she ate far too much and ended up with a terrible stomach ache. Ever since, she has associated the sight and taste of strawberry ice cream with sickness and vomiting. Even thinking about it makes her nauseous. It's another case of classical conditioning.

Something similar happens when you see a picture of George Clooney wearing an Omega watch and feel a sudden, powerful urge to buy that particular brand. Classical conditioning is a form of subconscious learning. My younger sister does not want what was once her favourite flavour of ice cream to make her feel sick, and nobody intends to be swayed by advertising. Yet it does and we are.

Operant conditioning is more conscious than classical conditioning. Instead of a dog that unconsciously starts to drool at the sound of a bell, think of one that is willing to sit, offer its paw or play dead in order to increase the likelihood of receiving a treat. If you give the dog the treat, it will probably perform the same trick again. On the other hand, if you scold it, the chances are that it will try something else. That annoying beep in your car when you forget to fasten your seat belt is a form of operant conditioning, because you learn and then remember what you have to do to make it stop. Therefore, operant conditioning requires conscious action.

Another form of learning is simpler than either classical or operant conditioning. Known as habituation, it involves no more than getting used to something. My first job was as a sales assistant in a clothes shop, and I would sometimes lock up in the evening without remembering to turn off the pounding music. I had become so used to hearing it throughout the day that I no longer noticed it.

The most complex type of learning involves absorbing information from others. You can't master playing a piano, driving a car or playing football solely through operant conditioning. The rules are too complicated. Your first driving lessons involve sitting in the back seat and watching your parents drive. Consequently, many years later, when you have your first official lesson, you have at least some idea of what to do. You learn to play football by watching games on TV, playing video games and having kickabouts with your friends. You observe other people, then try to emulate what they do.

The psychologist Albert Bandura demonstrated this process in a rather unpleasant experiment. He placed a child alone in a room, where they were shown a film of an adult beating up a clown doll. Later, when placed in a room with the doll, the child would usually start to hit it. Moreover, the likelihood of this response increased significantly if the film showed the adult receiving some sort of reward for beating up the clown.

Focused learning

Some types of learning – such as habituation and classical conditioning – never make their way into our conscious memory. We become accustomed to certain situations and anticipate future events subconsciously. When tackling more complex tasks, such as playing the piano or driving a car, we rely on the information we have previously stored in our data and motor memories. The data memory is the repository for information such as the rules of the road and how to interpret the symbols on a sheet of music. Then, through practice, we improve our skills, which are stored in the motor memory.

As anyone who has attempted to learn a new skill will tell you, certain pieces of advice and information sink in immediately, whereas others have to be endlessly repeated. Why does this happen? First and foremost, it's important to concentrate and remain focused, which requires a healthy thalamus (see Fig. 1) and fully functioning frontal lobes. However, some tricks can aid concentration. For instance, students are more likely to remember the details of a text that is written in an elaborate typeface rather than a simple typeface because the former forces them to concentrate.

You are more likely to remember what you've read or experienced if emotions such as fascination, joy or even anger are involved in the learning process, because all of these emotions enhance our attentiveness. The stored memories will then provide valuable information for the amygdala in the future (see Fig. 1). Equally, however, very strong emotions can have a detrimental effect on our ability to store important memories. Robbery victims who have stared down the barrel of a gun in terror tend to remember the weapon in great detail, but struggle to recall the clothes the robber wore, his height or even the colour of his skin.

Everything we remember enters the brain via one or more of our senses. All of the information is encoded in various areas of the cerebral cortex before the hippocampus combines everything into a single experience. It is also the hippocampus that links the new information to previously stored information. Once it has completed these complex tasks, the new experience is stored in the brain's long-term memory. We know that our long-term memories are stored in various parts of the cerebral cortex, but more research is needed into precisely where each type of information is stored.

The human memory is associative, meaning that it works better if we are able to link new information to something we have previously experienced or with which we are already familiar. If you can manage to link what you want to remember to something that means a lot to you, it will sink in properly. By contrast, if you try to remember something that you don't fully understand, you will probably forget it.

A well-known memory technique exploits this associative quality by linking new information to previously stored memories. In your mind, you imagine entering a house and then associate each new word you hear with a specific room as you mentally walk through the building. Mnemonic devices work in much the same way. You create a catchy rhyme or a funny word based on the first letters of whatever you want to remember. In school, we remembered how to spell the word 'necessary' by reciting, 'Never eat chips; eat salmon sandwiches and remain young'. Later, as a medical student, I remembered where each heart valve was located by silently repeating the phrase 'all patients trust me': A stood for aortic valve, P for pulmonary valve, T for tricuspid valve and M for mitral valve. Eventually, I became so familiar with the heart's valves that I no longer needed the mnemonic, but it certainly helped in the early stages of the learning process.

The more you repeat something, the better you will remember it, but this process is not the same as building up your muscles by working out at the gym. Although mnemonics are useful for remembering specific pieces of information, they do not lead to a superior memory.

Storage

Each of our long-term memories is not stored neatly in a single mental drawer that can be opened as and when we need to access it. The visual information relating to a particular memory is stored in the visual cortex, the aural information in the cortex for hearing, the emotions in the amygdala and the information derived from touch in the somatosensory cortex. Of course, we remember our own pain and try to avoid experiencing it again, but this might be reinforced by the memory of a TV programme in which someone trips over and involuntarily shouts, 'Ouch!' or the memory of seeing a boy cringe as he witnesses his friend falling off his bike. It's not just the sensory information that we remember, but how we felt when we saw, heard, smelled or touched it.

From dating to a permanent relationship

In laboratories all around the world, researchers are searching for an answer to the question: how is information stored in the brain? In other words, what happens when we press our mental 'store' button?

As mentioned in Chapter 1, each of us has around 86 billion neurons, which is certainly a lot. On the other hand, very few parts of the brain have the ability to make any more. For instance, when we first study algebra at about the age of ten, the brain does not suddenly start producing 'algebra cells' to store all of the new information we receive. Instead, we have to repurpose our existing neurons, all of which are already full of earlier memories.

Everything we think, learn and remember is sent as a series of electrical and chemical signals via an interconnected neural network. Each electrical signal is sent through the neuron cell body and into the nerve fibre, or axon. At the tip of each axon, this signal is converted into a chemical signal that is sent across the so-called giant synaptic cleft, which in reality is a not-so-gigantic twenty nanometres across. So the nerve cells are not in direct contact with one another: they are separated by a gap of 0.00002 millimetres. On the far side of this gap, the chemical signal arrives at the next neuron in the network. The signal is transmitted through a synapse and converted into another electrical signal, which then speeds towards the next nerve cell in the chain.

The more synapses you have, the easier it is to adapt to new challenges. So how do you increase your number of synapses? Simple: learn something new! And that doesn't mean you have to take a course in Latin or study for a degree in philosophy. Learning how to play table tennis or joining a salsa class can be just as effective. As your number of synapses increases, your neurons will be able to form ever more neural networks. However, a word of warning: if you don't practise what you've learned, the synapses you've created will start to disappear. In other words, the brain constantly generates and destroys synapses, although those that we use regularly become permanent. Meanwhile, those that we use all the time are reinforced by a mysterious substance called LTP.

Mr LTP

You know you're a nerd when you feel star-struck upon meeting an eighty-year-old professor. I was that nerd when I met Terje Lømo – the Norwegian physician who discovered LTP.

Each neuron has between 10,000 and 15,000 points of contact – synapses – with other neurons. However, these synapses are not all equally efficient. LTP stands for long-term potentiation, meaning that the synapse becomes more efficient as it is used over time. It develops when neurons send signals to one another so often that their sensitivity gradually increases. It's rather like building a friendship: the neurons that communicate with each other frequently across the synapses become more closely linked. Eventually, it's as if neuron number two is more attentive every time neuron number one starts chatting: 'Your signals are really faint, but I can hear them and I'll pass them on. But only because it's you!'

Professor Lømo discovered LTP back in 1966, but it was a long time before the rest of the scientific community understood the crucial role it plays in learning. Our synapses learn, so the neural networks that we use most often become

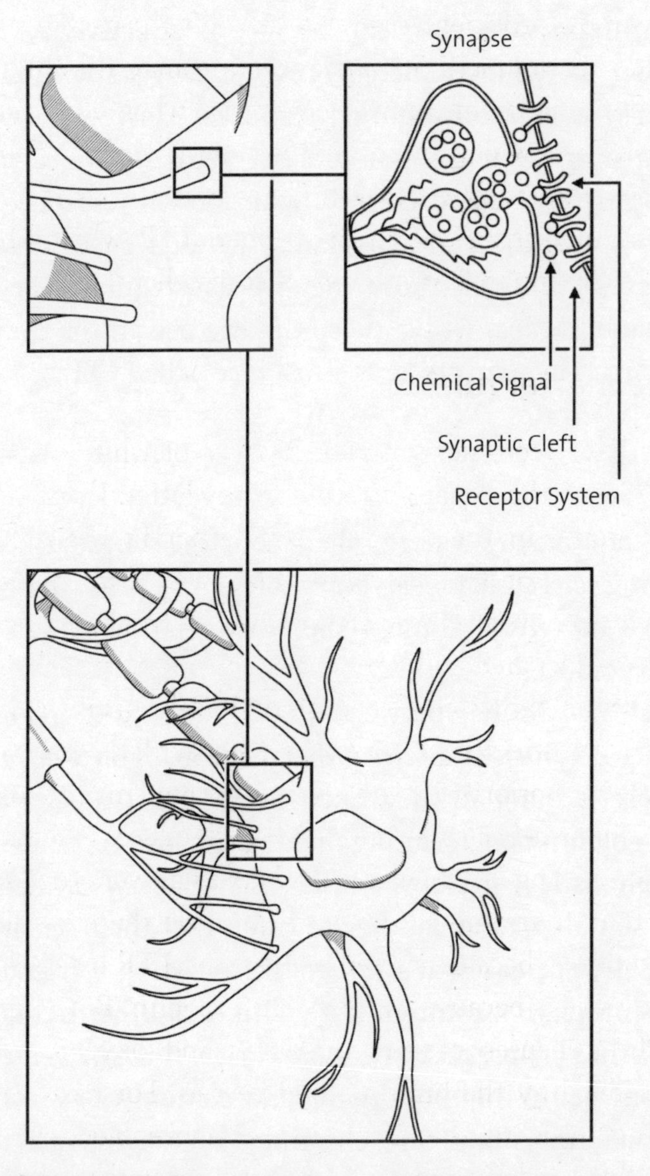

Figure 10: The large diagram shows an axon from one neuron contacting the axon from the next neuron in a network. The place where the information is transmitted from one cell to the next is called the synapse, and the gap between the two axons is called the synaptic cleft. These are shown in the upper-right diagram. The information is transmitted when the first neuron releases a chemical signal (neurotransmitter) that acts on the next neuron's receptor system.

increasingly proficient over time. You've probably experienced this yourself. For instance, you may have noticed that your movements are very awkward when you first try to master a new dance move. However, if you continue to practise, it starts to get easier. That's partially due to your neurons increasing and utilising their LTP, which allows them to communicate better with each other.

White matter rules!

As we have seen, nerve tissue consists of white and grey matter, with the synapses found in the latter. Information is not stored in the synapses but rather in networks of neurons, each of which includes highways of white-matter axons that allow electrical signals to pass from A to B. The axons are insulated with myelin, which not only helps them transmit the signals more efficiently but also gives the white matter its distinctive colour. Some particularly important pathways are prioritised and given additional myelin, which results in ultra-fast communication and reduces the risk of a message getting lost along the way. In other words, the most important neural networks not only have the most sensitive synapses because of their increased LTP but also the best highways because of their extra myelin. Both myelin and the synapses require nutrients and oxygen, which are supplied by the brain's blood vessels. For this reason, learning leads to the formation of more blood vessels, which are needed to meet the neural networks' increased energy demands.

Although our knowledge of the formation of new synapses, thicker myelin around the axons, new blood vessels and the role of LTP is increasing all the time, we don't yet fully understand the vital processes of learning and memory.

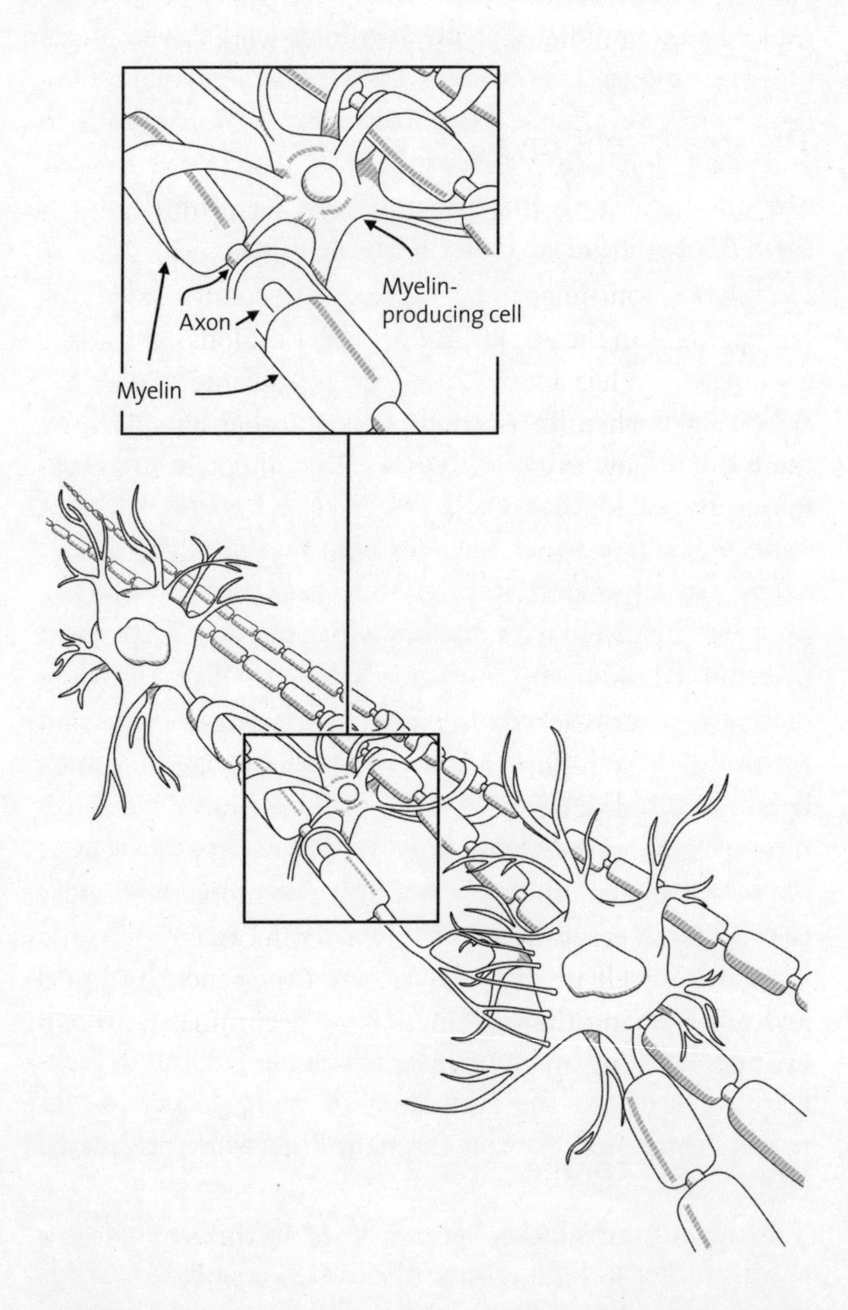

Figure 11: Axons are insulated with myelin, which allows the electrical signals to move faster.

Nevertheless, these discoveries are helping us reach a better understanding of how the human brain works.

The 10 per cent myth

The idea that we utilise only 10 per cent of our brain is a myth that has proved difficult to refute, especially as Hollywood continues to perpetuate it in countless movies. For example, in the 2014 film *Lucy*, Scarlett Johansson plays a woman who has a new kind of drug implanted under her skin. As the chemical starts to leak into her bloodstream, Lucy learns how to exploit her brain's full potential to dramatic effect.

From a neuroscience perspective, this is nonsense. Nine-tenths of the human brain does not lie fallow in the absence of chemical stimulation. In fact, we use every one of our neurons. If we didn't, evolution would have ensured that the brain never reached its current size, because its demand for energy is so high. However, that's not to say that there is no untapped potential, because our neurons could form thousands more *networks* than they do today. Moreover, as we have seen, the synapses become more efficient as their LTP increases. In this way, the brain can organise and reorganise itself in response to new experiences and new learning, storing the information we accumulate through practice, training and education.

Unlimited storage capacity

The brain is not a locked hard drive from the moment you are born. Its 86 billion neurons are constantly changing, so you can always learn more and improve your skills. Memory storage never definitively ends; rather, it's an ongoing process, with new experiences and memories continually merging with older memories.

If you're not concentrating, you will store very little. And if you're tired after long days of revision in the run-up to a big exam, you might feel as if your head is full and that there simply isn't room for anything else. However, many researchers now believe that our mental storage capacity is almost unlimited. If you forget something, it's not because you've deleted it from your hard drive; rather, you're just having trouble retrieving it. Have you ever tried to remember a name that won't come to mind? Then, a few hours later, it pops into your head when you're focused on something else entirely? Researchers say this proves that the memory was not deleted; it was just temporarily difficult to access. With this in mind, we know that our brain performs a sorting process, both consciously and unconsciously. It evaluates what's important and what isn't, and details that it deems insignificant are rarely stored.

On the other hand, it's hard to say anything with certainty in this field. After all, memory is a flexible process. Researchers now believe that some seemingly unimportant details are stored at least temporarily, presumably because the brain suspects that they might turn out to be useful in the future. For example, you might surprise yourself by being able to remember the colour of the car that drove past your house right before your burglar alarm went off.

Remembering

Our ability to recall a specific memory rests on the stability and strength of the neural networks where it is stored. As we have seen, our neural networks are strengthened by frequent use, and a strong memory is more easily remembered. However, the act of remembering is also a creative process in which you blend new and old memories rather than simply recall one specific experience. Moreover, as fragments of memory are stored in various parts of the cerebral cortex, they need to be remembered as fragments before being combined into a meaningful whole.

Both your surroundings and your mood can help you recall certain memories. We've all entered a room only to find ourselves unable to remember why we walked in there. So we return to our starting point and immediately remember what we had intended to do. In this case, our surroundings help us remember. Similarly, it's easier to recall memories from last year's trip to the mountains if you're currently in the mountains. And you're more likely to remember pleasant memories when you're happy, and sad memories when you're feeling down.

When asked to remember a list of words, most people are able to recall those at the beginning and end but struggle with those in the middle. However, if hints are provided, almost everyone can recall the words that they failed to remember initially. This points to the fact that the words haven't completely vanished from our memory; we just need some help to access them.

When asked a question on a particular subject, we know immediately if we know nothing about it. We don't need to scroll through our memory to reach that conclusion. Similarly, we know instantly if we know something about it, although, in this instance, we will probably have to think for a while to remember the information we need to answer the question, and this can vary depending on how long it's been since we last retrieved it.

We retrieve memories in two distinct ways: we can actively recall them or we can recognise them. The latter involves comparing something we see or hear with the memory of it. Our brains even have a special facial recognition region. This allows you to identify your father effortlessly in a crowd of two thousand men even though you'd never be able to describe him with sufficient precision for a stranger to do the same. Recognition, then, is a much more passive process than recollection. It's as if something just clicks without any need to think about it.

Familiar faces – such as those of family, friends or even favourite celebrities – generate activity in specific areas of the brain whenever we see them. For example, you probably have a Jennifer Aniston neuron! When the signal to a particular neuron was measured with an electrode in a group of patients who were waiting for operations to treat their epilepsy, the scientists discovered that the neuron reacted whenever Jennifer's picture came up, regardless of whether it was a close-up or a long shot, from a movie or real life, or any other variable.

When you need to remember something specific, your brain reactivates the neural network that was used when the memory was formed. Fortunately, the two sensations are not identical, as that would feel like hallucinating every time you thought about a previous experience. Instead, your brain tells you that you are picturing a memory and also reminds you where you are at present.

We know that our memories help us make future choices. But we don't need to remember absolutely everything we've ever experienced in order to achieve that. So it seems that some of the episodes that are stored in our long-term memory eventually form a general knowledge database that allows us to generalise on the basis of previous experience even though we lose the ability to access them as individual memories.

How to improve your memory

By now, you should understand the importance of concentration when trying to store new information. Your powers of concentration – and therefore your memory – are adversely affected by both sleep deprivation and stress, so you're unlikely to benefit from worrying about an upcoming exam or revising long into the night. If your stress increases as a big event – such as an exam – approaches, it's extra important to focus on what you need to remember well in advance. And if you can manage to engage with the material and link it to your emotions, so much the better. Similarly, the more senses you employ, the better something will stick. For instance, if you read a passage from a textbook out loud, your brain receives information from both your eyes and your ears, which often creates a stronger memory. But don't overdo it because this technique works best if you limit yourself to reciting only the most important passages. Then you should try to repeat all of the essential information from memory. In other words, practise your retrieval skills, and maybe even correct your memory when it introduces errors.

Since repetition helps memory, allow me to reiterate the point: make the time to listen to yourself, or read aloud to others. Test yourself, review sample exam questions or have friends ask you questions about the subject you are studying. Practising retrieving knowledge from your memory is far more effective than reading silently through the same material time and again. Your memory will improve as you work actively with the material you have already learned. Remember, you don't just need to store information well; you also need to become adept at retrieving it.

It's also advisable to cut down on the wine before trying

to commit something to memory, at least if you expect to be sober when you need to access it. If you learn something while you're drunk, you'll remember it better when you're intoxicated. This is because the retrieval process works more smoothly when our current situation is similar to the one when we learned the information. So, if you know you're going to take an exam in a silent auditorium, you should revise in a quiet environment. Language has an important role to play here, too. Bilingual speakers of Russian and English who live in the United States find it easier to remember details from their childhoods if they discuss them in Russian rather than English. Our memory is also better if we see a mental image in colour rather than black and white.

So it is possible to optimise the functioning of almost any memory. However, all memories are not the same. Some people can remember an amazing amount of detail after the briefest flight over a large city; others can recall entire phone books. Yet, such people usually lack some of the basic skills that the rest of us take for granted. That's because their exceptional memories are the result of certain types of brain injury. We have no definitive explanation for why a person's memory can improve after their brain is damaged, although there are plenty of theories. One of these focuses on injuries or impairments in the left half of the brain, which usually plays a vital role in filtering information.

People who have the paradoxical combination of intellectual disability and super-developed memory are known as savants. About fifty savants from around the world have been studied in various research projects. One of them, Kim Peek, learned to read before he could walk. He had an enlarged head, no corpus callosum between the left and right hemispheres, and no cerebellum. Understandably, he was diagnosed as suffering from a mental disability, yet he could read two pages of a book at the same time – one with each eye – and remember everything in great detail. For ever! Ultimately, he was able to recite the contents of twelve thousand books. In 1984, the screenwriter Barry Morrow met Peek and based his script for the Oscar-winning film *Rain Man* on his life.

From the kitchen window where I grew up, I could see a tree that was usually populated by all sorts of different birds. I would watch the tree for hours and learned to recognise bullfinches, great tits, house sparrows and the Eurasian jay. I remember the jay especially well, because it has such beautiful blue feathers. Coincidentally, this species often crops up in discussions of memory because it hides food for the winter in branches, under tree roots and in all manner of cracks and crevices. The Eurasian jay is not considered particularly intelligent, even by bird brain standards, yet studies have shown that it can remember the locations of several hundred of these caches.

When we were in elementary school, we thought the smartest kids in the class were those who could memorise the most capital cities. The truth is rather different: you can bone up on a lot of facts, but you can't bone up your intelligence. Henry Molaison (H. M.) had no recollection of what had happened yesterday, yet he was personable and intelligent. Kim Peek could read a large book in an hour and remember every word, but he couldn't button up his own shirt.

Using your nose to remember

And soon, mechanically, weary after a dull day with the prospect of a depressing morrow, I raised to my lips a spoonful of the tea in which I had soaked a morsel of the cake. No sooner had the warm liquid, and the crumbs with it, touched my palate than a shudder ran through my whole body, and I stopped, intent upon the extraordinary changes that were taking place. An exquisite pleasure had invaded my senses, but individual, detached, with no suggestion of its origin ... I was conscious that it was connected with the taste of tea and cake, but that it infinitely transcended those savours, could not, indeed, be of the same nature as theirs ... Will it ultimately reach the clear surface of my consciousness, this memory, this old, dead moment which the magnetism of an identical moment has travelled so far to importune, to disturb, to raise up out of the very depths of my being? I cannot tell. Now that I feel nothing, it has stopped, has perhaps gone down again into its darkness, from which who can say whether it will ever rise? Ten times over I must essay the task, must lean down over the abyss ... And suddenly the memory returns ... The sight of the little madeleine had recalled nothing to my mind before I tasted it ... But when from a long-distant past nothing subsists, after the people are dead, after the things are broken and scattered, still, alone, more fragile, but with more vitality, more unsubstantial, more persistent, more faithful, the smell and taste of things remain poised a long time, like souls, ready to remind us, waiting and hoping for their moment, amid the ruins of all the rest; and bear unfaltering, in the tiny and almost impalpable drop of their essence, the vast structure of recollection ... [T]he whole of Combray and of its surroundings, taking their proper shapes and growing solid, sprang into being, town and gardens alike, from my cup of tea.

Marcel Proust, *Remembrance of Things Past*

Have you ever noticed that a smell or a taste can evoke a powerful memory? The parts of the cerebral cortex that are most associated with memory and olfactory information are located right next to each other. And they are closely linked functionally as well as anatomically. Thus, a familiar smell can help us recall a previous experience. This is known as the 'Proust phenomenon'.

Most of the information that ultimately arrives at the hippocampus has already visited several other parts of the cerebral cortex. These areas make associations between various pieces of information and interpret any new messages that come in from the senses. By contrast, all of the information relating to smells takes a shortcut from the olfactory cortex straight to the hippocampus, without making any detours through the cerebral cortex's associational areas. Olfactory data doesn't even enter the thalamus, in contrast to all other sensory information. The shortcut is useful because olfactory axons are both uninsulated and very narrow, which makes their transmission of electrical signals extremely slow.

The olfactory cortex is also closely linked to the amygdala, the almond-shaped group of neurons that plays a key role in stimulating and controlling our emotions. Consequently, the memory of a specific smell will generally trigger a particular emotion. If the memories you associate with smells seem unusually powerful, genuine and important, that's because they are emotionally charged.

The olfactory nerves are the only neurons in the whole central nervous system that are exposed to the open air – at the top of the nose. They detect dozens of smells that we are able to recognise immediately, although we have trouble describing them. For example, how would you describe the aroma of a strawberry to someone who's never smelled one? Would you be able to explain it with sufficient precision that they would recognise it when they first encountered it? Probably not. Yet, once you've stored a scent in your memory, you don't forget it, because your olfactory memory is amazingly stable.

Blacking out

'Blacking out' is not a scientific expression, but it's used frequently, especially when describing the consequences of drinking too much alcohol. To be sure, it requires a lot of alcohol to reach such a point, but the brain can become so incapacitated that it is no longer able to store memories. And that will mean you won't remember anything.

A more controversial topic is the concept of 'suppressed memories', which supposedly cannot be recalled after a person suffers some sort of trauma. At present, the evidence both for and against the subconscious suppression of memories following a traumatic incident is inconclusive. However, it is generally accepted that we can *consciously* repress some memories. In 2007, researchers at the University of Colorado showed a series of unpleasant images to a group of volunteers and found that the subjects were able to exercise a certain amount of control over their memories of the pictures. The conclusion was that the subjects were able to stop the retrieval process by actively forcing themselves not to remember the images.

Of course, before we can suppress something unconsciously or repress something consciously, first we need to create a memory. And traumatic experiences are usually burned into our memory. As a rule, we usually remember them extremely well.

Dementia is brain failure

Increasing forgetfulness is a normal part of the ageing process, because the neurons in older brains lose some of their connections and start to die. Indeed, it is possible to see the shrinkage in human brains over time in standard CT scans. The hippocampus, which plays one of the most important roles in memory, is one of the first areas to decline with age.

When the kidneys fail, we call the condition kidney failure; when the heart fails, we call it heart failure; when the liver fails, we call it liver failure. Yet, for some reason, when the brain fails, we call it dementia. Translated directly from the Latin, dementia means 'out of one's mind', which is a fairly accurate description of what happens to someone who suffers from the condition. But a better term might be 'brain failure'.

Dementia is divided into many subgroups, based on the specific part of the brain where the failure originates, but ultimately the condition spreads to such an extent that it's difficult to tell the various groups apart. Alzheimer's disease – the most common form – is linked to the presence of a particular protein, which in turn damages the brain's neurons. The condition seems to begin in the temporal lobes, right next to the hippocampus, so, inevitably, the short-term memory is affected in the very early stages of the disease. At this point, the patient's personality and sense of humour will be almost unaltered, but they will start to forget to turn off the oven, blow out a candle before going to bed or buy the item they need from the supermarket. Initially, it's possible to compensate for the deterioration by drawing up lists, but eventually that's not enough. This is often the point when the sufferer – or a close relative – realises something is wrong and consults a doctor.

My great-grandmother developed Alzheimer's, and there is one episode that I remember particularly well. One weekend, she spent hours in the kitchen, preparing a fantastic meal for all the family. But no one turned up, which really hurt her feelings. Later, it transpired that she had forgotten to invite any guests.

Dementia sufferers often feel a sense of grief, especially in the early phase, when the rest of the brain is working properly and understand that something is going badly wrong. Later, once they have lost most of their working memory but while the information in their long-term memory remains intact, they can experience something of a second childhood. As the disease spreads, however, the long-term memory starts to disappear, too. And the personality. And the sense of humour. In short, relatives are forced to watch, helpless, as their loved one gradually fades away.

Fortunately, some important discoveries have already been made on the road towards solving the Alzheimer's mystery. For instance, a group at Stanford University discovered that if elderly mice receive blood transfusions from young mice, they start to produce more neurons in the hippocampus. So maybe youthful blood contains a factor that has the power to limit and possibly even reverse the forgetfulness that comes with old age? If we can manage to supplement such findings with establishing why harmful proteins start to accumulate as we age, we will be well on the way to developing a viable treatment and arresting the disease's progression.

The second most common form of dementia is known as vascular dementia. This disease involves a narrowing of the brain's blood vessels, which results in mini-strokes – or transient ischemic attacks (TIAs) – and the death of the affected areas' oxygen- and nutrient-starved neurons.

Therefore, this type of dementia usually does not develop gradually; rather, it progresses intermittently, depending on when and where the strokes occur. As with all other diseases that are linked to blood vessels, the principal risk factors are an unhealthy diet and insufficient exercise.

The other types of dementia don't affect the memory first; instead, at least in their early stages, they cause personality changes and hallucinations. Eventually, though, they impact on the memory, too.

According to a report by the Alzheimer's Society, 850,000 people were living with dementia in the UK in 2015. And that figure is likely to double by 2050. As yet, there's no cure. So, what can we do to reduce the risk? Obviously, it's hard to do anything about the ageing process, but we know that a well-trained brain is more able to resist the deterioration that is associated with Alzheimer's. If you continue to exercise your brain well into old age, it will tolerate higher levels of the protein that seems to play a key role in memory loss and personality change. Consequently, while sufferers who keep their brains active still deteriorate, their symptoms develop at a slower rate. The way to combat vascular dementia is even simpler: eat well and get a lot of physical as well as mental exercise.

Mr Appelsine

When I was a toddler in Norway – long before I started to study English – my mother told the family a story about an English teacher who would never admit to making a mistake or not knowing the answer to a question. During one of his lessons, he anglicised the Norwegian word for orange – calling it an 'appelsine' – and wouldn't admit that he'd got it wrong when one of his students tried to correct him. From that moment on, the whole school called him Mr Appelsine.

Everyone else laughed, but I didn't get the joke as I didn't speak English, so my mother had to explain why it was funny. Now that I do speak English, I have to concentrate in order to avoid making the same mistake. Whenever I'm about to request a drink on a plane, I invariably find myself on the verge of asking for an 'appelsine' before repeating 'orange, orange, orange' quietly to myself. I don't want to acquire the nickname Mrs Appelsine!

There should be an easier way to unlearn knowledge, but our brains have no 'delete' button. Indeed, the more we try to forget something, the more we remember it. Every time I'm about to blurt out the wrong word for orange and need to correct myself, I'm forced to access exactly the same neural network as I did when I stored my mother's story about Mr Appelsine in the first place. As a result, my memory of hearing that story grows stronger and stronger, so I'll probably remember it for the rest of my life.

False memories

Although the memories we access on a regular basis tend to be clearer than others, the human memory should never be considered entirely reliable. This is because we store information in the form of a 'memory skeleton' that consists of only the most important details. When we retrieve it later, we use our general knowledge and understanding to flesh it out on the basis of a series of assumptions and other memories. Of course, this embellishment has the potential to introduce numerous errors, yet our brains are desperate to put some meat on the bones of the original memory skeleton. Studies have shown that we are highly receptive to suggestions that help us fill in the gaps in our memories both when we construct them and when we retrieve them. There are many examples of so-called false memories, when witnesses are unconsciously influenced by leading questions or media coverage and change their testimony as a result.

Many memories must be recalled and stored again numerous times before they find a permanent place in the long-term memory. However, the entire memory can be changed during this re-storage process: the relative strengths of the neural connections can be altered and the memory might be associated with new sensations, environmental conditions, expectations or knowledge.

Elizabeth Loftus is a talented scientist who has dedicated much of her life to studying false memories. For instance, she has proved that terminology plays a key role in how we remember particular incidents. In one experiment, she showed the same short film of a car accident to two groups, one of which was told that the car was 'smashed', while the other was told that it was 'hit'. The first group remembered seeing much more broken glass than the second group, even though all of the subjects witnessed the same incident. In short, Loftus's description of the event had a profound effect on what they remembered seeing with their own eyes.

Celebrate your forgetfulness

With no memory, we wouldn't recognise our family or friends. We wouldn't even recognise ourselves. So it's hardly surprising that a lot of people wish that they could remember more than they do. But be careful what you wish for. If you have an average memory, you should probably be satisfied with that. After all, your brain is able to sift through all of your experiences, pick out whatever's important and store it away for future reference, and discard the insignificant details. In short, your memory works like a filter that protects you from the sensory overload that you encounter every day.

A handful of people are cursed with being unable to forget a single thing that has happened to them. I'm not talking about the memory champions who employ a range of strategies to remember dozens of packs of playing cards and get their names in the *Guinness World Records*. And I'm not talking about savants. Even Kim Peek could not compete with the amazing memory of an American woman named Jill Price. You can pick any date at random and she is able to rattle off what the weather was like, exactly what she and the people around her were doing, and what was on the news. She remembers everything, down to the tiniest, most inconsequential detail, and likens her memory to a movie that never stops. Every day, she sees the present and the past simultaneously, as if on a split screen. Understandably, although most people call her extraordinary memory a gift – the technical term is hyperthymesia – Jill herself describes it as a burden.

So, you should start to celebrate your forgetfulness. Although most of our memories are not high-resolution images of the past, they don't need to be. They are good enough to help us make wise decisions in the future, and that's always been the human memory's most important task. It's what makes it such an important component of the superstar brain.

Chapter 4

The Brain's GPS

A rat runs contentedly around its large cage, searching for the chocolate chip a researcher slips through the bars at regular intervals. The rat is wearing a device that resembles a hat with wires emanating from it. The wires record every occasion when a specific neuron in the rat's temporal lobe sends a signal. At first glance, the neuron seems to be sending totally random signals, but gradually a pattern starts to emerge when the rat's position at each moment of neural activity is plotted on a grid. If you draw straight lines between the points where that neuron sparks into life, you end up with a series of geometrically perfect, interconnecting hexagons: every side of every hexagon is exactly equal in length (see Fig. 12). For many years, video-game developers have believed that a hexagonal grid is far preferable to a square grid when creating a virtual universe. Now it turns out that the brain was millions of years ahead of them.

Grids in your brain

This groundbreaking discovery was made by a team of Norwegian brain researchers led by May-Britt and Edvard Moser in 2005. They called the neurons that generate these hexagonal patterns 'grid cells', and later showed that our sense of orientation consists of multiple grid-cell modules, each of which operates at a specific scale. We use larger-scale modules for large areas where details are unimportant, and finer scales for small areas where we need high resolution. The researchers located all of these modules in the region of the cortex right next to the hippocampus in the temporal lobe. The scale – which ranges from small at one end of this cortical region to enormous at the other – increases by the square root of two from one module to the next.

You are here

In the past, most people used paper maps to find their way around. In the pre-GPS era, you had to flip the map over and rotate it in order to orient yourself. Sometimes you would have to search for landmarks like mountains or churches to figure out where you were. Wouldn't it have been great if there had been a little red dot on every map that said 'You are here' every time you looked at it? Well, your brain actually has something like that.

Less than ten years after the Mosers discovered grid cells, they shared the Nobel Prize in medicine with US-British researcher John O'Keefe, who discovered neurons he named 'place cells'. These are the cells that enable us to know where we are at any given time: in effect, they are our brain's little red 'You are here' dots. Having fitted his rats with similar hats to those used by the Mosers, O'Keefe measured the neural activity in the hippocampus and found that some of its neurons emitted strong signals whenever the rats were in a specific location in their cage, yet remained inactive when they were anywhere else.

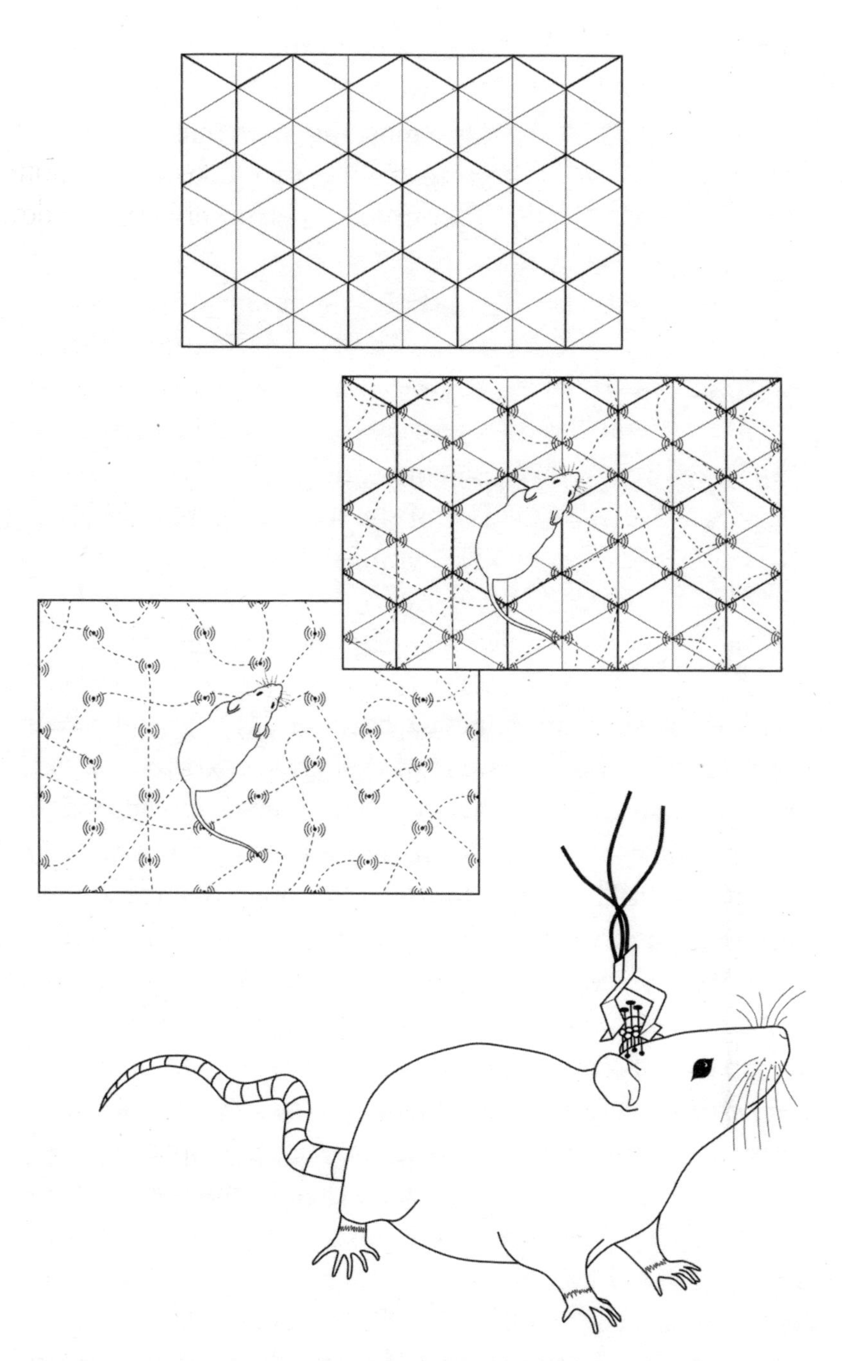

Figure 12: A rat's neuron sends signals at specific moments to create a grid of perfect hexagons.

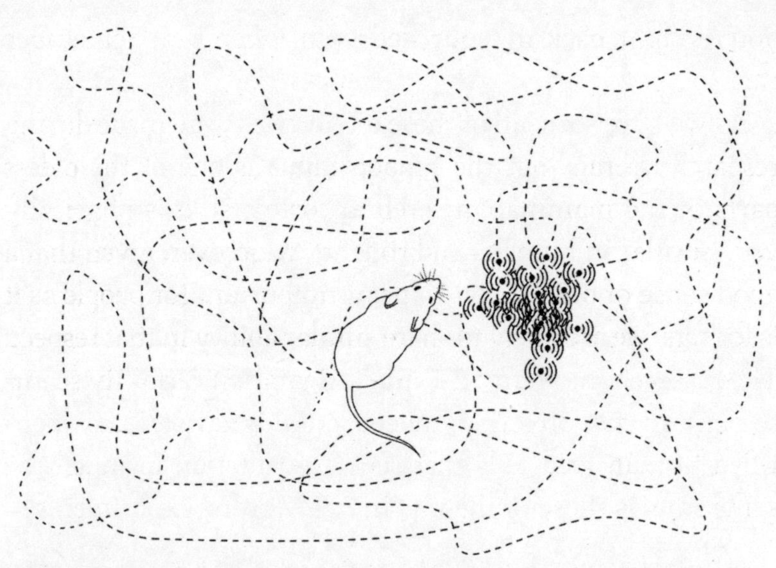

Figure 13: The place cell emits a signal whenever the rat visits a specific part of its cage.

When a surgeon removed both of Henry Molaison's hippocampi as well as some of the adjacent cortex, he lost both his place cells and his grid cells. Hence, in addition to being unable to recognise his nurses after the operation, he couldn't find his way to the bathroom. Unsurprisingly, place cells and spatial orientation are closely connected to memory. Indeed, the vast majority of our memories are linked to the specific places where they were created. Based on further studies of rats, it seems that place cells are not concerned solely with where we are right now; they also provide information about the memories that are associated with particular geographical locations. For instance, in all likelihood, the place cell that emitted a signal whenever you opened the toy box in your childhood bedroom now sends a clear signal each time you reminisce about playing there, even if you're a thousand miles away when the memory comes to mind. In other words, mentally if not spatially, you're right back in your bedroom when you remember that toy box.

As we have seen, all of these discoveries were made during research on rats, but the hippocampus is one of the oldest parts of the mammalian cerebral cortex, so it is physically very similar in humans and rodents. Moreover, given that a good sense of orientation is just as important for people as it is for rats, we are likely to share similar ability in that respect. The research into human spatial orientation certainly seems to point in that direction. Indeed, studies have already identified human grid cells that seem to function in much the same way as those of the rats in the Mosers' experiments.

Map and compass

I'm the first to admit that I've never been too good at finding my way around. As a result, I often just follow along passively, which means that I don't practise and improve my orientation skills. Every once in a while, though, I am absolutely certain about which way to go. For instance, ten years ago, I visited Budapest with a couple of friends. On one occasion, I was so convinced that I knew which way to go that I found it almost painful when one of my friends insisted we should go in the opposite direction. Based on our respective track records, there was every reason to believe that she was right and I was wrong, but we were in no hurry, so she humoured me and accompanied me in the direction I wanted to go. She pointed out landmarks along the way and patiently explained why they meant we were getting ever further from our intended destination. Eventually, through this process of gentle persuasion, she managed to recalibrate my feeble orientation centre. We did an about-turn, walked back to our starting point, then headed in the direction she knew we should have taken from the beginning. By now, though, my head direction cells weren't screaming in protest.

In many ways, our head direction cells act like a compass, although they don't tell us which way is north, south, east or west. This is because they aren't linked to the earth's magnetic poles, but to our own balance organ in the inner ear. A particular head direction cell is activated whenever you turn your head in its direction, regardless of whether you're standing on your hands or have your eyes closed. However, if you keep your eyes closed for a long time, the signals from your head direction cells become less precise. Studies on rats have shown that if the lights are turned on and off repeatedly, the rat becomes disoriented and its whole head direction system temporarily breaks down. Moreover, if the rat is placed in new surroundings for only a couple of minutes each time, the head direction cells seem less able to orient themselves on the basis of landmarks. This causes

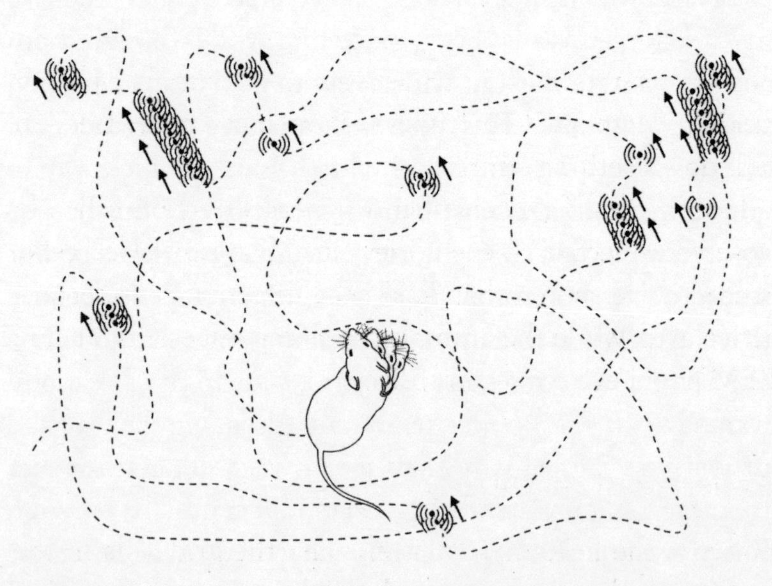

Figure 14: A rat's head direction cell emits a signal when its head is facing in a particular direction, regardless of the direction in which it is moving.

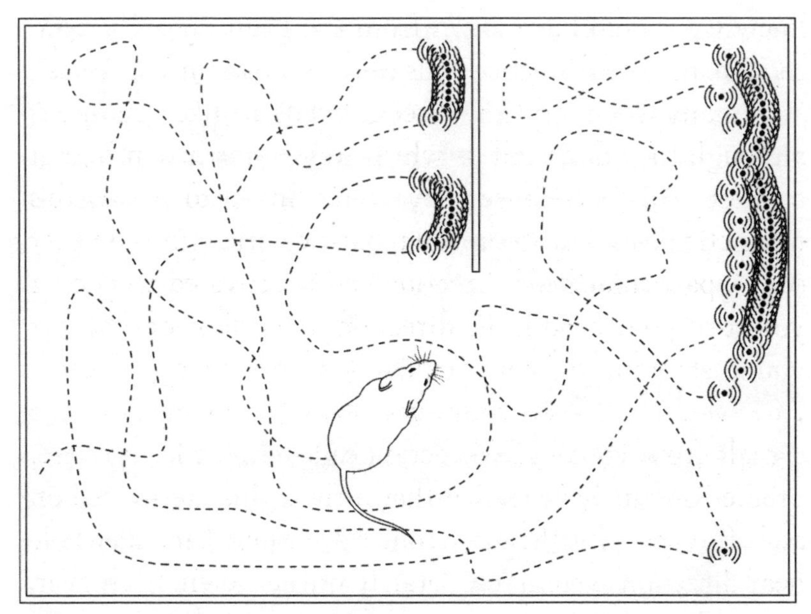

Figure 15: Border cells identify any boundary, be it the exterior of a rat's cage or an interior dividing wall. Here, a specific neuron emits a signal whenever the rat approaches a border to its right. A different border cell will perform the same function whenever it approaches a border to the left.

them to start sending inconsistent – or even completely random – signals. I sometimes wonder if my head direction cells do something similar.

It seems highly likely that our head direction cells are closely connected to memories, just like our place cells. Studies have shown that rats' head direction cells remain active even while the animals are asleep, especially during REM (rapid eye movement) sleep.

To here, but no further

Between the head direction cells and the grid cells in the section of the cortex around the hippocampus, another small group of cells tells you when you are approaching a boundary or a border, such as a mountain, a wall or a fence. However, they emit a signal only when you are *right next to* that border. So, if a rat's cage is enlarged between two experiments, the border cells remain inactive until it reaches the new edge of the cage. They have an important role to play in telling the place cells and the grid cells where to focus their attention.

Fred Flintstone's car

Fred Flintstone, the cartoon caveman, has a car with no motor and stone wheels. He has to run to make the car move, with his legs poking through the floor. May-Britt and Edvard Moser constructed a similar car in miniature for their rats. The rats would propel it to a chocolate reward at the far end of a four-metre track. If they were allowed to run freely to the chocolate, they could achieve speeds of 50 centimetres per second. However, the Mosers chose to limit their speed to 7, 14, 21 or 28 centimetres per second by increasing the friction on the car's wheels (see Fig. 16). The point of the experiment was to measure the activity in hundreds of neurons while the rats ran towards their rewards. After amassing results at various speeds, the Mosers identified dedicated speed cells: that is, neurons that emit signals depending on how fast the rat is running. These speed cells send their signals independently of landmarks and irrespective of whether the rat runs in daylight or complete darkness. Interestingly, if a rat is allowed to run at its own pace, its speed cells give a clear indication of the speed it is *about* to achieve, as opposed to its current speed.

While the head direction cells tell the grid cells the direction in which a rat is moving, the speed cells tell the rat how fast it is moving (or will be moving). The grid cells

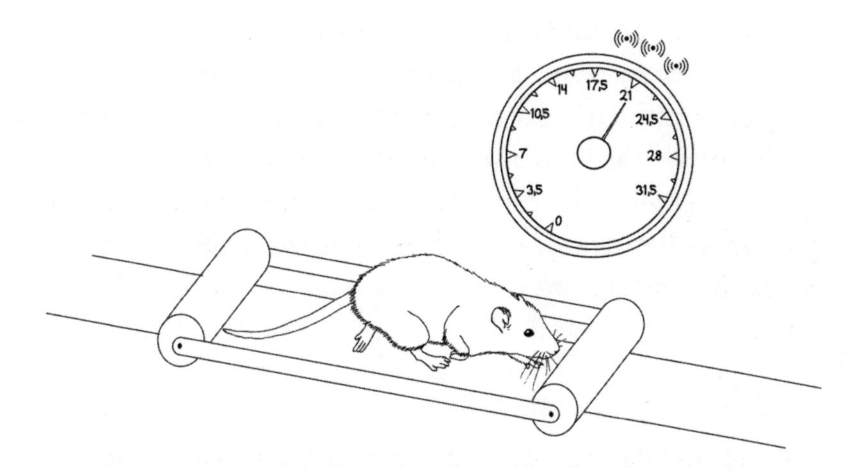

Figure 16: Rat in a Fred Flintstone car with a speed cell emitting a signal at 21 centimetres per second.

know how to utilise this information, and the rat's grid is updated. In short, the grid cells draw a map, the border cells indicate its boundaries, the place cells work out where we are within those boundaries, and the speed cells tell us how long it will be before we arrive at one of them. All of these neurons contribute in their own distinctive ways to our spatial orientation. Taken together, they are the brain's global positioning system – its GPS – which comprises a speedometer, a compass and boundary markings.

It's not just the temporal lobe

As we have seen, the place cells are located in the hippocampus, while the grid cells are in the adjacent cortex. Thus, as far as we know, two crucial components of our spatial orientation system are found exclusively in the temporal lobe. Head direction cells, however, are located not only around the hippocampus but in some other parts of the cortex as well, and in the thalamus and the basal nuclei (see Figs. 1, 8 and 9).

To orient ourselves in our surroundings, we need more than a mental map, a compass and a speedometer. We need our occipital lobe to process visual information and recognise landmarks. And we need our sense of touch and an awareness of our own motion, such as when our foot touches the ground. Both the parietal lobe and the cerebellum have roles to play in this awareness. In other words, when we move, our ability to orient ourselves doesn't rest solely on our perception of landmarks on the horizon. Our brain also receives constant information about how we're moving and where our arms and legs are positioned at any given time. It's by processing all of this information that we are able to navigate and orient ourselves effectively.

A fully functioning parietal lobe combines visual data with information from the other senses. Rats with damaged parietal lobes can still move around with assistance from their cerebellum. However, their interpretation of sensory information is clearly impaired, which means they struggle to find hidden food and their way back to their own cage. And it's not just rats that lose their way after suffering parietal lobe injuries: humans who have had strokes in their parietal lobes struggle to identify the right path even in familiar surroundings.

Are men better than women at finding their way around?

No – in fact, the opposite may well be true. At present, though, the only firm conclusion we can draw from thousands of hours of research is that women and men employ different – not necessarily better or worse – strategies when finding their way around. This has generated a series of seemingly contradictory results depending on how different groups of researchers have designed their experiments. For instance, several studies on virtual orientation – an important aspect of many computer games – have shown that men outperform women. However, critics of these experiments have suggested that the findings simply reflect the fact that the male subjects have probably played more video games throughout their lives than their female counterparts. In other words, they've had more practice at virtual orientation. When it comes to orientation on actual terrain, the sexes record almost identical results. However, as a group, women seem to rely more on landmarks, such as hills, church spires and other conspicuous physical features, whereas men are more inclined to use their sense of direction. This is also evident when they give directions. A woman will typically say, 'Take a left at the grocery store, then go straight until the road starts to turn to the right,' whereas a man would be much more likely to give instructions to drive east, west, north or south. Multiple studies have shown that, in general, the women's strategy of utilising landmarks is better than the men's at helping them find their way back to their starting point after visiting a new place.

All of these studies are based on averages, though. Obviously, some women are far better than the average man, but others are far worse. Unfortunately, I'm definitely in the latter camp. And I can't even use the excuse that I was 'born like this'. Of course, some people are born with a better sense of direction than others, but, as we know, our brains are plastic, so spatial orientation can be improved with practice. If you spend all your time thinking, 'I'm terrible at this,' 'I'm sure to get lost,' or 'If I go by myself, I'll never arrive on time,' then it quickly becomes a self-fulfilling prophecy. There's a tendency for women to have little confidence in their spatial skills. Maybe that's why the myth of male superiority is so deeply entrenched. Self-confidence is always an important contributory factor in performance. A study published in the journal *Science* in 2006 showed that women who were told that men were naturally better at maths did worse on arithmetic tests than women who were told that the sexes performed equally well.

The cab driver's brain workout

So, can you train your hippocampus and develop a better sense of direction? When researchers at University College London decided to explore this question, they didn't have to look far for perfect subjects. London's streets are a real hodgepodge – the city has very little of the geometrical urban planning of Paris or New York. So, London's cabbies have to remember a complex labyrinth of 25,000 roads, not to mention thousands of tourist attractions and other important locations. Most trainee taxi drivers spend two to four years learning the routes before they feel sufficiently confident to take the test and gain their licence to drive a black cab. Even then, the failure rate is close to 50 per cent.

When the University College researchers scanned the brains of a group of taxi drivers and a group of people of equivalent age and IQ, they found that the cabbies' hippocampi were significantly larger than those of the control group. Was the extra size due to the cabbies' years of practice and subsequent experience on London's streets, or did the rigorous selection procedure simply weed out any applicant with a naturally small hippocampus? Further analysis pointed towards the former conclusion, because the researchers found that cabbies with many years' experience had larger hippocampi than those who had just started their driving careers. Later, the same research team followed a group of aspiring cab drivers from the start of their training to the moment when they passed the exam, scanning their brains at regular intervals. The hippocampi of those who passed the exam grew significantly larger during the course of their training due to the formation of new neural connections, and possibly through the creation of new neurons, too.

The hippocampus is one of the few locations in the brain where new nerve cells can form, and the London study is one of the clearest indicators that our experiences can have a physical impact on our brains.

How can you improve your sense of direction?

London's taxi drivers have the ability to picture a map of the city in their heads and calculate the shortest route to a destination. If they simply entered an address into a GPS and passively followed the system's instructions, their hippocampi would not grow during their training and throughout their careers.

Similarly, if we use landmarks to orient ourselves and thus construct a mental map, we use our brain more actively than if we just follow the instructions on a GPS screen. If you take the same route home from work every day, your brain is much more passive than if you explore a new route. As we have seen, neural connections deteriorate when they aren't used. If we automatically follow the GPS when it tells us to go straight ahead for 200 metres then turn right, we are not exercising any of the neural connections in our hippocampus. We are not noting the locations of any of the landmarks we pass or putting them into a coherent context because we have been staring at a screen the whole time. Hence, in addition to depriving ourselves of the pleasure of seeing a beautiful old church or a blossoming cherry tree, we lose much of the geographic and cultural context of the world around us. Our brains would be much better served if we used our own sense of direction or even followed an old paper map.

Japanese researchers asked three groups of subjects to find their way on foot to a specific destination. The first group used mobile phones equipped with GPS software, while the second made their way to the same destination using a paper map. The researchers gave the members of the third group a route to follow, and they were not allowed to use any additional navigational aids. Unsurprisingly, the first group was the worst at explaining the route they'd selected, and they struggled to draw a map of it after reaching their destination. Rather more surprisingly, they also ended up walking further than the other two groups and making more stops en route. The group that had to navigate on their own after the route was explained to them did the best. Of course, GPS devices can be time-savers, but don't forget that you have your own internal GPS and it's really quite good at what it does.

If you want to keep your navigational skills in top shape (and you don't have anyone to tell you the best route to take), then a paper or even a digital map is preferable to GPS. The size of GPS screens often prevents you from seeing where you are and where you're going simultaneously, and a researcher called Véronique Bohbot believes that this might have serious long-term consequences. She claims that an overreliance on GPS devices may make our brains so passive that we will be much more likely to develop diseases such as Alzheimer's later in life. In short, while London's cabbies have taught us that the hippocampus is physically enlarged through extra use, Bohbot insists that our increasing dependence on GPS is making it smaller. Alzheimer's affects the neurons in the hippocampus in the early stages of the disease, but a fit and healthy hippocampus will probably tolerate more damage before the person starts showing pronounced symptoms.

We should be happy that we don't need to rely on our mobile phones and sat navs to find our way around. As we have seen, the brain's internal GPS can be just as effective in helping us reach our destination. It's especially useful when we need to find our way in new places, or when navigating to the kitchen in the middle of the night. Without this ability, we would walk around in a series of concentric circles and never be able to work out which way to go.

CHAPTER 5

THE EMOTIONAL BRAIN

Imagine how boring the world would be without happiness, love, disappointment or anger. It's important to understand your own and other people's emotions. When you recognise your own emotions and understand why you're experiencing them, you can learn how to make a few of them take a detour around your cerebral cortex in the future. This will enable you to nip an emotional outburst in the bud before you hurt or offend anyone.

Strong emotions can make people shave their heads, attack the paparazzi with an umbrella, grab the microphone from a teenage singer who is giving an acceptance speech or scream childish – and premature – celebrations in the immediate aftermath of an election. In such circumstances, carefully cultivated images can be destroyed in an instant.

All of us wish we had more control over our emotions. We all want to be like the doctor who allows an outburst from a frustrated patient to play itself out before calmly explaining the best course of action. By contrast, no one wants to be the lawyer who starts crying in frustration when a vital witness changes their story under cross-examination. That sort of behaviour seems so ... unprofessional. Yet sometimes our emotions seem to control us, rather than the other way around.

There are two different pathways for expressing emotions. One involves a detour through the cerebral cortex, which allows it to talk some sense into the more primitive parts of the brain: 'There's no reason to be afraid of slow-worms. They're not poisonous.' In my case, however, whenever I see a slow-worm, the visual information does not take a trip through my cerebral cortex. Consequently, my body reacts as if it's on the brink of death, even though, rationally, I know that's not the case. Even if I remind myself on the way to the zoo's reptile house that all of the exhibits are safely locked in their cages, and none of them can hurt me, these reassuring thoughts evaporate the second I catch my first glimpse of a slow-worm. The primitive parts of my brain send my whole body into a state of high alert before my cerebral cortex has a chance to talk any sense into me.

The pathways that various sights, sounds and other sensory information take in the brain vary from person to person, and within each person. For example, I have no problem flinging myself out of an airplane as long as I've got a parachute on my back, or off a bridge with a bungee cord attached to my leg. Yet the sight of a legless lizard, even if it's on TV, invariably freaks me out. I'm not predestined to suffer from this for ever: the cerebral cortex can be forced to take charge in cases where a phobia has previously enjoyed free rein. However, the opposite can happen, too: one frightening episode can be sufficient to generate intense fear of something similar in the future. And the cerebral cortex doesn't always calm you down. Sometimes it will warn you to steer clear of something that doesn't seem too threatening at first glance, such as the seemingly friendly stranger who came to the playground and offered you sweets when you were a child.

In most cases, we seek a second opinion from the cerebral cortex before acting on our emotions. Otherwise, we would all find ourselves jumping up and down on Oprah Winfrey's sofa and screaming about how great it is to be in love. Nevertheless, in some situations, we should be grateful that our feelings occasionally take a more direct route. If a car is careering towards you, you won't have time to contemplate what's going on, who's driving the car, what their intention might be, or whether you'll rip your new coat if you jump out of the way. In such circumstances, it's advisable just to hurl yourself over the nearest hedge, rather than mull over your options.

The only place where it's acceptable for adults to scream out loud is at a sporting event. For instance, most parents try to keep the swearing to a minimum when their child accidentally drops a bag of flour and covers the whole kitchen in a thin, white dusting. Instead, you take a deep breath, count to ten, then clean up. A large proportion of everyday life involves reining in our emotions in this way. On the other hand, what would the world be like without strong emotions? What if we didn't feel guilty when we lied, didn't experience love for our children, partner, family and friends, or lacked the motivation that drives us to achieve a challenging goal? The truth is that we are all utterly dependent on our feelings, both positive and negative. They guide us through our lives and help us advance a little further each day.

Feeling with your brain

Whenever the cerebral cortex reaches a decision, the body's hormones and autonomic nervous system follow up on it.

If the cortex is the ruler, the autonomic nervous system – which consists of the sympathetic nervous system (which turns you on) and the parasympathetic nervous system (which turns you off) – is one of its subjects. The sympathetic nervous system causes you to tremble the first time you write on a whiteboard in front of a class and makes your hands sweat when you are asked a difficult question, but it also helps you react quickly and escape from danger, if necessary. As soon as the danger is over, the parasympathetic nervous system calms you down, brings your heart rate back down to normal and slows your breathing. While your sympathetic nervous system channels most of your blood to your muscles in preparation for fighting or fleeing, your parasympathetic nervous system redirects it back to its routine functions, such as powering the digestion of food in your intestines. Without the sympathetic nervous system, we wouldn't focus our full and undivided attention on staying on an icy road after an unexpected skid. Without the parasympathetic nervous system, we would continue to have butterflies in the stomach and remain in a state of high alert even after driving home safely.

Of course, the brain controls the body and its reactions to specific situations. Its most primitive part starts the process which causes the adrenal glands to start producing the stress hormone adrenaline. For instance, adrenaline is released when someone you've been attracted to for a long time kisses you, when you feel anger or when you're frightened. Once it has entered the bloodstream, it causes an increase in heart rate, respiration rate and blood pressure.

In most situations, the cerebral cortex knows precisely which emotions we are feeling at any given time. But a team of US researchers wanted to explore what we feel during physiological arousal, so they injected two groups of volunteers with adrenaline. The first group was informed of the hormone's effects – such as a pounding heart and rapid breathing – which meant the volunteers had a logical explanation for the symptoms they experienced after the injections. None of them reported any change in their mood. The other group was not informed of the effects of adrenaline, so they started searching for explanations for why they suddenly felt so frantic. If an actor joined the group and started behaving erratically, the volunteers tended to relate their racing heartbeats and sweaty hands to irritation at his behaviour. In other words, their brains identified the most logical cause of their physical symptoms. Thus, it appears that arousal due to adrenaline – or other neurotransmitters (chemical messengers) – can contribute to an emotional change, with the brain deciding what kind of emotion we are experiencing. It's not the hormone that makes you feel angry or happy; rather, your cerebral cortex interprets what you're feeling after analysing the situation.

When you're in love, your brain emits signals that make your heart beat faster and cause you to focus all of your attention on the point where your lover's hand is resting on your thigh. However, the sensation of being in love is not located in your thigh or your heart; it's in your brain. And the same is true of every other emotion. It's difficult to pinpoint precisely where each one is found, but some patterns are emerging. A number of structures that lie deep within the brain – around the corpus callosum on each side – comprise the limbic system, and this is generally considered to be the seat of all of our emotions (see Fig. 17). As we saw in Chapter 3, one of these structures – the hippocampus – converts information from the working memory into long-term memories, and those memories tend to be stronger when they're associated with strong emotions.

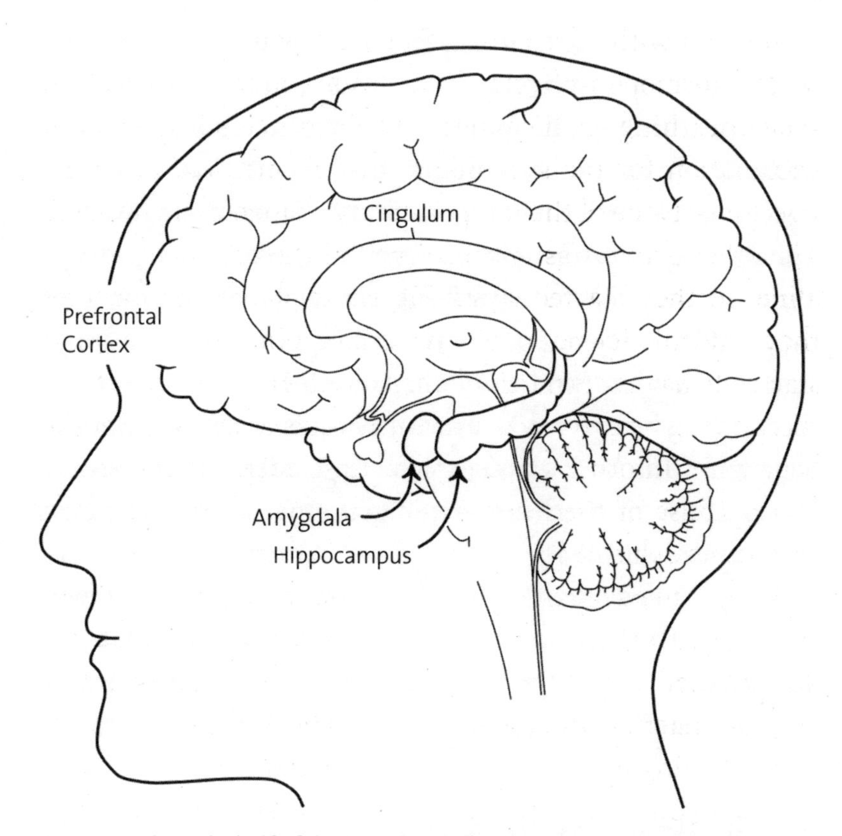

Figure 17: The right half of the brain, seen from the middle, along with the left temporal lobe's hippocampus and amygdala. Important areas for emotions include the region of the cerebral cortex known as the cingulum, the hippocampus and the amygdala, all of which are parts of the limbic system. A more recent part of the human brain – the prefrontal cortex – has the capacity to override emotions.

Just as hunger motivates us to find food, emotions motivate us to address needs such as safety and relationships. This is the essence of evolution: survival and reproduction.

Everything mental is also physical, because all of our emotions are controlled by various chemical substances in the brain. For instance, when you interpret someone's behaviour as friendly or generous, you experience a surge of dopamine, serotonin and oxytocin. All three of these neurotransmitters make us feel good and motivate us to try to produce the same pleasant sensation in the future.

As in this instance, numerous chemical substances tend to be active in the brain at any given time. Some of them aid communication from one neuron to another, whereas others affect many neurons simultaneously in a particular area. Because so many different neurotransmitters are working in unison, your brain is able to adjust your mood and emotions to find the best balance for your particular situation. Moreover, it doesn't only control which emotions you feel, but how strongly you feel them. So it decides when a sensation escalates from grumpy to furious or from nervous to terrified.

We all need emotions to function in society, but permanently heightened emotions or emotions that are experienced at inappropriate times or in response to the wrong stimulus are not good for our health.

Smiling your way to happiness

When you smile, your brain receives signals from your facial muscles that improve your mood. People who were asked to smile while watching a cartoon found it funnier than those who were asked to furrow their brows. When you make an angry facial expression, the anger and fear centre in your brain – the amygdala – is activated.

Botox is poisonous to neurons, so when it's injected into a patient's forehead, the muscles stop working because the neurons that control them are no longer able to send and receive signals. Consequently, the patient's amygdala receives no information from the paralysed facial muscles. Researchers studied the impact of Botox injections on patients who were treated for frown lines caused by contractions of the small depressor muscles near their eyebrows. The results were astonishing: 90 per cent of the subjects in the study group who had been profoundly depressed for at least six months prior to their Botox treatment were depression-free within two months of undergoing the procedure.

Botox is certainly not an established or recommended treatment for depression, but this study raises an interesting question: is it harder to feel down when you're unable to frown?

Bad moods are bad for you

... and good moods are good for you. Of course, life is never that simple. Our mood is governed by far more complex mechanisms than our facial expressions. Nevertheless, people who claim that a bad mood is all in your head are right. Everyone has experienced a bad mood at some point in their life. There's always *something* that will make you feel sad: a tear-jerker movie, betrayal, disappointment or grief, for instance. But once the situation has changed and a little time has passed, most of us are able to drag ourselves out of it.

Depression is very different. It isn't on the normal spectrum of emotions; instead, it's a disease that has a profound effect on the way some people think, behave and view the world. It extends well beyond sadness, and it is rarely triggered by one specific event. It robs the sufferer of energy, motivation and the ability to experience happiness, exhilaration, satisfaction or meaning in their everyday life. People who suffer from depression also have shorter lifespans than those who never experience it. There are many reasons for this. Sufferers tend to isolate themselves, which makes it harder for them to access help when it is needed; they neglect other aspects of their health; and we know that chronic stress harms both the body and the brain. Therefore, depression should not be viewed as a *mental* illness. A change in mood is linked to a series of *physical* changes: in the brain's chemistry; in the areas of the brain that are active; in which neural connections continue to function; and in which neurotransmitters the neurons release.

Studies of depression have focused much of their attention on the neurotransmitter serotonin, which helps most people maintain a sense of equanimity and optimism. Normally, it is released into the cleft between two neurons and then received by the second neuron's receptor system (see Fig. 10). Multiple studies have found that deeply depressed individuals have fewer receptor systems to capture the serotonin, which seems to indicate that physical changes in the brain's neural networks contribute to depression. Under normal circumstances, any excess serotonin in the synaptic gap between two neurons would be reabsorbed by the neuron that released it. However, a group of medications known as selective serotonin reuptake inhibitors (SSRIs) blocks this reabsorption process, which means that the surplus serotonin remains in the synaptic gap for longer. This gives the second neuron's receptor system more of an opportunity to absorb it and, ultimately, pass it on. The end result is that people with too few receptors can achieve relatively normal levels of serotonin transmission, which makes life much more enjoyable. Hence SSRIs have acquired the nickname 'happy pills'.

Unfortunately, they don't work for everyone, because depression is not a single disease. Rather, it's an umbrella term for multiple conditions that share similar symptoms, and we don't know enough about the chemistry behind those conditions to provide targeted treatment. Maybe in the future we'll be able to take a scan of the brain to establish how many serotonin receptors there are in various regions, and that will help us predict whether SSRIs will be effective. At present, though, it's more a case of trial and error.

When depressed people have relatively normal levels of serotonin transmission, problems relating to another neurotransmitter – dopamine – might be to blame for their low mood. You tend to feel sad if your brain is unable to absorb this hormone correctly, even after experiencing something that should have been a pleasurable experience.

Sufferers from Parkinson's disease generally do not receive a diagnosis until they start to display a number of physical symptoms, such as a resting tremor in the hands and difficulty initiating movement. However, we've known for years that other symptoms, including depression and a deteriorating sense of smell, often precede these physical manifestations of the disease. This is hardly surprising because Parkinson's kills not only the neurons that produce dopamine but also those that transmit it from the brainstem to the basal nuclei, which makes sufferers far more susceptible to problems relating to personal motivation and low mood as well as muscle control.

Yet not all Parkinson's patients are depressed. Indeed, recent research suggests that fewer than half – 45 per cent – suffer from the condition. The depressed patients seem to have fewer dopamine receptors in their limbic systems than those who manage to maintain a positive outlook, so medications that increase the availability of this neurotransmitter are proving effective at alleviating not just their muscle control issues but also their depression. This is consistent with the findings of earlier experiments which found that inhibiting dopamine in the midbrains of mice causes symptoms of depression, whereas boosting dopamine transmission reduces depression.

Although we know that depression in Parkinson's patients is due to specific physical changes in their brains, we should not focus all of our attention on the neurons that produce and transmit dopamine. Both conversation therapy and learned strategies for tackling negative thoughts, in addition to medication, can also initiate physical changes in the brain. Moreover, they can help sufferers of depression to mitigate the accompanying chronic stress that can prove very harmful over the long term.

The brain's green-eyed monster

If you become green with envy, an area in your cerebral cortex between the right and left hemispheres – popularly known as the 'jealousy spot' – fires up. Jealousy is caused by our fear of losing something we value. When the subjects in an experiment read glowing reports about the achievements of a group of eminent people, they felt jealousy and their jealousy spots were activated. By contrast, when the same subjects read about VIPs suffering some sort of misfortune, they experienced Schadenfreude – a sort of malicious pleasure – and activity was detected in a particular region of their basal nuclei.

Sex on the brain

Simply stimulating the cerebral cortex in the cleft between the two halves of the brain is sufficient to cause a monkey to have an erection. However, a complete, fulfilling sexual experience involves activity in almost every part of the brain at one time or another. For instance, a man's occipital lobe plays a pivotal role when he looks at a plunging neckline, and a woman's is active when she sees a tight t-shirt stretched over a well-defined torso. If you were then to place your hand on that body part, the signals from your palm and fingers would travel to the parietal lobe on the opposite side of your brain. Meanwhile, your frontal lobe dictates what you find attractive in the first place, with some help from your limbic system. Thus, it causes you to focus much of your attention on people who possess the features that appeal to you and disregard those who don't.

On the other hand, the frontal lobe is one of only two parts of the brain (the other is the amygdala) that remains inactive during an orgasm. The deactivation of the frontal lobe makes perfect sense, since its passivity prevents the person in question from mulling over the possible consequences of what they're doing. We do not yet fully understand why the amygdala, which is usually involved in primitive emotions, is also deactivated. However, researchers have observed hypersexuality and uncritical sexual behaviour in patients with certain types of brain injury, so perhaps the amygdala needs to be shut down at critical moments to suppress the development of these harmful conditions. For instance, an injury to the inside of the temporal lobe, where both the hippocampus and the amygdala are situated, can result in Klüver–Bucy syndrome. Like most syndromes, it takes its name from the people who first described it – in this case, the German-US psychologist Heinrich Klüver and the American neurosurgeon Paul Bucy. Sufferers have significant memory problems, including an inability to store new memories. They are also unable to feel fear or anger. However, they continue to experience strong – and sometimes abnormal – sexual urges.

As a teenager in New Jersey, 'Kevin' suffered from epilepsy, but neurosurgeons managed to eliminate the attacks by removing the part of his brain that was causing them. For a time, the procedure enabled Kevin to live a normal life: he was happily married and enjoyed his job. Everyone found him personable and he was a well-liked member of his local community. However, a few years later, his epilepsy returned, so he decided to undergo another operation. Once again, the procedure eradicated his seizures, but this time there was a serious complication: Kevin lost his inhibitions. Sometimes this manifested itself in relatively harmless ways: for example, he would play the same song on the piano for nine hours at a stretch. But he also developed an insatiable appetite for food . . . and for sex. In addition to downloading hours of regular pornography, he developed a fascination for movies and pictures involving very young children.

At his trial for possession of child pornography, Kevin's defence rested on the claim that *he* hadn't committed the offence; his surgically altered brain was to blame. By then, he had been diagnosed with Klüver–Bucy syndrome. The judge took this into account when passing sentence.

All of our brains emit signals that turn on sexual desire, but most of us also generate signals that help us to rein it in. Keeping up appearances and maintaining self-control in the immediate vicinity of a ripped torso or perky breasts involves more than just our temporal lobes. The region of the cerebral cortex between the two halves of the brain (the cingulum) and the prefrontal cortex (see Fig. 17) are actively involved in dampening our passion, too. The sweet, elderly grandmother who used to be interested in nothing but her vegetable garden yet now pinches male nurses' bottoms may have suffered damage to one of these areas, usually as a result of frontal lobe dementia.

To do or not to do

Procrastinators put off what they have previously decided to do, even if this makes them feel guilty, experience stress and, obviously, achieve less than they should. Procrastination is a side effect of how we assess various tasks. It's not that you can't do a particular task; you simply lack the motivation to begin. Of course, you still intend to do it; just not today. You choose short-term pleasure over long-term benefit.

There is a pattern to the tasks that we decide to shelve. Theoretical tasks demand more self-control than physical ones, and repetitive tasks require more self-control than varied ones. Therefore, we're more likely to do the weeding in the garden than fill out our tax returns. This is particularly true if the deadline for submission of your tax return is months away. The longer the deadline, the less attractive working on that project becomes. And the harder we expect a task to be, the more likely we are to procrastinate.

If you tend to shelve tasks that seem just too daunting, try dividing them into a series of smaller goals. If a physical job is more tempting than a theoretical one, reward yourself by using it as a break in the course of getting on with the theoretical task. Before you know it, you'll have finished both. Most importantly, allow yourself to be a dreamer. People often postpone important tasks because they know they won't see any sort of payoff for a long time. By contrast, they are willing to tackle jobs that provide immediate benefits. So, try to visualise the hefty refund that will eventually come your way if you submit your tax return on time, or the praise that you'll receive if you give a stellar presentation. Don't worry about being full of yourself, just dream big.

Your brain is to blame when you procrastinate, but you

also have it to thank when you're motivated to get on with something. The way in which the signals travel between the neurons in your brain determines whether you'll stick to your New Year's resolution or whether you'll decide to hit the snooze button in the morning. Nobody is born a snoozer. Some neural networks deteriorate over time while new ones form when we learn. So, you can think your way to physical changes in your brain.

People who seem to have a limitless capacity for hard work tend to have more of the 'reward neurotransmitter' – dopamine – in their basal nuclei and prefrontal cortex than those who are inclined to procrastinate. Both of these parts of the brain have important roles to play in generating motivation. Healthy rats with plenty of dopamine choose to work for good food rather than eat bad food that is supplied to them regardless of the effort they put in. On the other hand, if the dopamine signals in their basal nuclei and prefrontal cortex are blocked, they make do with whatever is thrown into the cage. In other words, dopamine motivates us to work towards a positive outcome (or avoid a negative outcome), so the term 'reward neurotransmitter' is a little misleading as it does some of its most important work *before* we receive a reward, not afterwards.

Dopamine's effectiveness rests not only on the *amount* we produce but also on it reaching the *right areas* of the brain. For instance, high levels of dopamine in a part of the basal nuclei called the nucleus accumbens enables us to predict an eventual reward if we behave in a certain way now. In short, the brain recognises that something important is happening and generates the necessary motivation to do something about it. By contrast, slackers have relatively low levels of dopamine in their frontal lobes and basal nuclei, but higher levels in their insula – the part of the cerebral cortex that lies behind the temporal lobe (see Fig. 5). If you find yourself lazily surfing the internet instead of getting down to important work, then you should try to increase the dopamine levels in the regions of your brain that are important for motivation. You can do this by linking your dopamine response to the achievement of specific goals. Give yourself a mental pat on the back every time you accomplish something important. Dopamine will flow as a result.

Be warned, though, this can require a lot of effort. The will to win isn't worth a damn if you're not prepared to put in the necessary legwork. Whatever the weather, champion marathon runners pound the streets every day to give themselves even the slightest chance of selection for the next Olympics. Sometimes the solution to low motivation is simply good old-fashioned determination and perseverance. There are times in life when you must be prepared to stick with something you find mind-numbingly boring or physically taxing in order to reap a reward in the distant future.

Angry winners

Evolution ensures that those members of a species with beneficial traits survive and pass on their genes to generations of descendants. So why, after millions of years of human evolution, do we still have brains that lose their temper? In what way is this emotion – which is often viewed in a negative light – an evolutionary advantage?

The answer is that anger helps us to keep unacceptable, antisocial behaviour at manageable levels because people who are tempted to act like jerks know that they will soon feel the wrath of those around them. While sadness and fear make us shy away from uncomfortable situations, anger drives us to resolve them. It is generated by neurons in the gyrus (in the middle of the brain, just above the corpus callosum), the cingulum and the left frontal lobe, all of which burst into life whenever you tell someone off for elbowing their way to the front of a queue or failing to show up for work on time (see Fig. 17). Interestingly, strong men and beautiful women – that is, those with obvious evolutionary advantages – often have hotter tempers than the rest of us, too. They're also adept at resolving conflicts to their own advantage.

A Dutch study found that it usually pays to express anger or irritation during negotiations. That's because we tend to concede more when sitting across the table from someone who's angry as opposed to someone who seems satisfied. We find their anger so uncomfortable that we back down time and again in a bid to soothe them, which means they secure a much better deal than if they'd remained calm.

Stress kills neurons

If your life is in imminent danger, digesting your breakfast or making some new white blood cells suddenly doesn't seem so important. So those routine functions are put on hold while your brain ensures that it and your muscles receive all the energy they need.

The instant your brain realises that you're under threat, it sends nerve impulses down your spinal cord to your adrenal glands and asks them to release the hormone adrenaline. As soon as this enters the bloodstream, your heart rate and blood pressure both increase and you start to breathe faster. All of this enables you to send oxygen- and nutrient-rich blood to your muscles and brain in double-quick

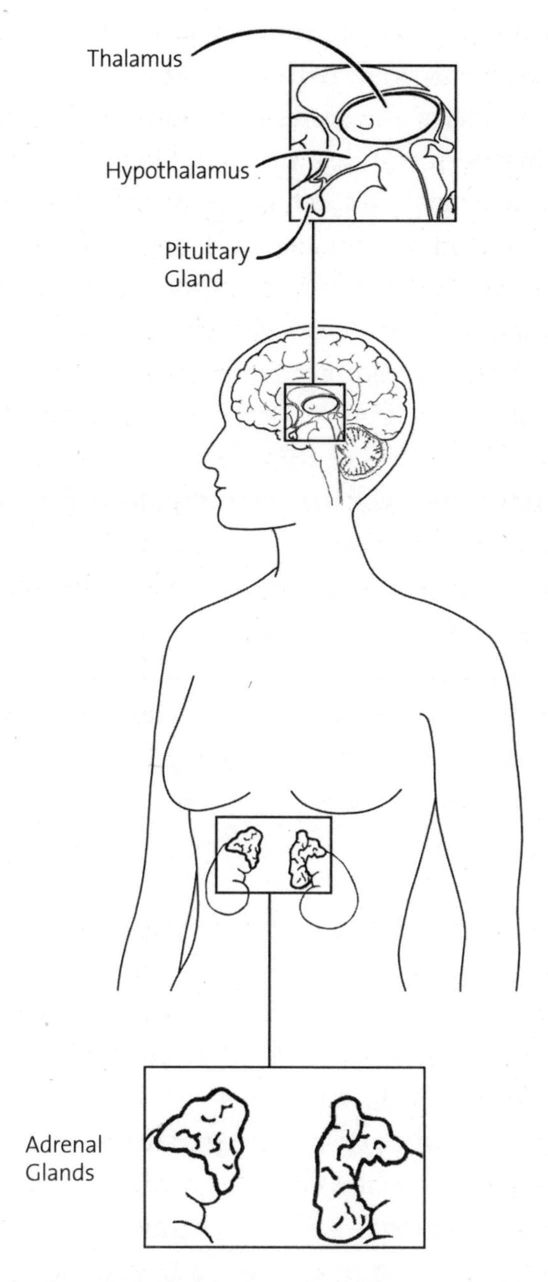

Figure 18: The right half of the brain shown from the middle, with a close-up of the three structures that are responsible for controlling the body's hormonal system. The brain's stress hormones stimulate the adrenal glands, which in turn release their own stress hormones into the bloodstream.

time. Meanwhile, your liver raises your blood sugar level in preparation for action. Without this impressive stress response, our species would never have survived on the African savannah.

The hypothalamus sits directly beneath the brain's gossipmonger, the thalamus. It controls the pituitary gland, which looks like a tiny pair of testicles hanging beneath the brain. After receiving an order from the hypothalamus, the pituitary gland releases a hormone that causes the adrenal glands to release cortisol (yet another stress hormone). This is vital during your stress response because it maintains high blood sugar and high blood pressure for as long as you need to escape danger.

There are multiple causes of stress – from everyday annoyances like long queues at the supermarket to major life events like the birth of a child or a natural disaster. Feeling stressed about a chemistry test helps you focus, set aside other projects and stick to learning the periodic table. Over the short term, then, a certain amount of stress is often beneficial. However, a stress response that continues for weeks or even years is almost always harmful. High blood pressure, especially when accompanied by high cholesterol and high blood sugar, increases the risk of a heart attack or stroke. One study showed that medical students had 20 per cent more cholesterol in their blood before an exam than after it. Accountants had both higher cholesterol levels and markedly faster blood clotting during stressful days in the run-up to the end of the tax year.

Continual stress doesn't just affect our cholesterol level, blood pressure and blood sugar. Cortisol also has a hand in accelerating the brain's ageing process. After travelling around the bloodstream, it reaches the hippocampus – the brain's memory centre. Once there, it helps us memorise stressful events – an essential survival mechanism as it enables us to avoid dangerous situations in the future. However, prolonged exposure to cortisol eventually damages and ultimately kills the neurons in the hippocampus.

It's impossible to avoid stress completely. We're all at risk of being flung into new, frightening or unforeseen situations. But we differ in terms of how long we let these new situations bother us and what we do to address them. Many studies have shown that people with a positive outlook on life live longer and happier lives than their more peevish neighbours. So don't let new or undesired situations stress you out for months on end. Calm down, remain positive, minimise your stress and start to enjoy life.

Anxious about anxiety

Early one morning, when I was sitting in the laboratory, one of my colleagues walked through the door and shouted, 'Hi.' This startled me because I hadn't expected anyone else to turn up for many hours and I was so focused on what I was doing that I hadn't heard his footsteps in the corridor. As a result, I dropped the glass cylinder I was holding and it shattered on the floor. My colleague wryly remarked that he wouldn't bother saying hello in future if that was my reaction.

In such situations, I'm tempted to curse my overactive amygdala – the brain's emotional centre. It makes me spill hot coffee every time I walk round a corner and meet someone unexpectedly, and I've got reams of notes in which a line of ink veers right off the edge of the page because someone shouted my name at an inopportune moment. Feeling startled is a completely spontaneous reaction. After all, the amygdala is one of the more primitive parts of the human brain, so it reacts to sensory information almost immediately. I simply don't have the time to interpret the abrupt sound I hear as a friendly greeting. Cecilie, one of my closest colleagues, understands this, so she now makes a discreet noise before approaching me. As a result, I've wrecked less of our research than I might have done!

There's a big difference between my sort of jumpiness and full-blown anxiety. Indeed, anyone who's experienced anxiety will tell you that it's one of the worst things a person can experience. In effect, your whole body reacts to the fact that your brain has just pushed the alarm button. You're so scared that your chest tightens and you feel a knot in the pit of your stomach. You're convinced that your heart is about to jump out of your ribcage. You become dizzy and feel as if you're about to faint. The effects are so severe that millions of people choose to isolate themselves and avoid places or situations that might trigger an attack. For example, if a person has suffered a panic attack in a grocery store, they might avoid all grocery stores or even refuse to set foot outside the house.

The amygdala sits at the tip of the hippocampus (see Fig. 17), and the two work very closely with each other. With the hippocampus's help, you are able to remember the last time you started to hyperventilate and almost fainted in the queue for the checkout. That memory alone is enough to activate your amygdala, so you become anxious about your own anxiety.

We feel anxiety when rational fear is allowed to run wild. In general, fear is a beneficial emotion: it stops us putting our hands in open fires and persuades us not to wander down dark alleys in crime-ridden parts of town. If we walk too close to the edge of a cliff, our amygdala decides that we should start to feel afraid so it sends out signals that make our legs shake and our palms sweat. In short, it protects us from going any further and doing ourselves an injury. As they say, forewarned is forearmed.

Unfortunately, sometimes the brain can be *too* wary. Some people feel their bodies gearing up to fight or take flight several times a day, even when there's no evidence of any danger. When this happens – due to the brain misinterpreting everyday situations and sensing a threat around every corner – we're no longer talking about regular fear, but anxiety. Most of the blood is directed to the body's major muscles, while the hands, feet and digestive system are all neglected. As a result, the body's extremities start to feel cold and numb and the mouth becomes dry. Meanwhile, the heart beats like crazy and the lungs start hyperventilating. This causes the arteries in the brain to contract, leading to dizziness and fainting.

While great advances have been made over recent years in the treatment of depression, anxiety is still treated with mind-numbing drugs that can lead to dependency and intoxication. However, there are alternatives. As we have seen, the brain is a dynamic organ: as you learn, it changes. Therefore, cognitive therapy focuses on explaining the symptoms of panic attacks, and especially why sufferers experience them. Armed with this information, a sufferer is less likely to spiral into terror the next time they start hyperventilating or feel their heart racing. Remember, the brain's frontal lobes do have the capacity to talk some sense into the more primitive limbic system. So, with a little practice, many people are able to take control of their anxiety and halt a panic attack before it's really had a chance to get going. In short, they learn to feel less anxious about their anxiety.

If thinking your way out of anxiety doesn't work,

physical exercise can generate new neurons and stimulate the release of several neurotransmitters that help reduce stress. Endurance training seems to be particularly effective in this regard. So regular exercise not only keeps you physically fit and healthy; it also helps you combat anxiety or depression.

Loving with your brain

Being in love makes your heart beat faster and your voice tremble. You feel like you need to pee about a thousand times before that first date with someone you've fancied for months. You have butterflies in your stomach and say that you love someone from the bottom of your heart. But all of these physical sensations are caused by activity in the brain, which then sends signals to the body.

As yet, we don't know what it is in the brain that makes us fall in love. However, we do know that love is an extremely complex emotion. It's not like fear and anger, which are almost entirely generated by the amygdala. When research subjects are shown pictures of people they love, several parts of the cerebral cortex, especially the insula, as well as deeper and more primitive parts of the brain, such as the basal nuclei and the limbic system, light up in their MRI scans. All of these regions of the brain are rich in the reward neurotransmitter, dopamine, which gives us the motivation to overcome our fear and approach people we find attractive in the first place.

Only 5 per cent of mammal species generally stick with one partner throughout their lives. We are one of them, and coyotes are another. Most coyotes are highly sensitive to the 'love hormone' – oxytocin – and tend to remain loyal to their partners. Those with brains that are less sensitive to it often go against type and switch partners. Most humans – as well as most coyotes – produce large amounts of this hormone during birth, nurturing and mating, and it certainly seems to play a role in strengthening the bonds of affection. For instance, studies have shown that men with naturally low levels of oxytocin are less likely to get married. But filling a nasal spray with oxytocin and squirting it up the nose of a sleeping partner with 'commitment issues' won't necessarily make him pop the question the next morning. Both the brain and love are much more complicated than that. Oxytocin is just one piece of the puzzle.

And it should be remembered that romantic love is just one aspect of this multifaceted emotion. For instance, parental love motivates us to dedicate years to caring for our children – a process that ensures the survival of our genes for at least one more generation. Brain scans have shown that this type of love is located specifically in the grey matter surrounding parts of the drainage system for the cerebrospinal fluid in the brainstem.

My daughter chose to come into this world almost two and a half months early. She was placed in an incubator and received all of the help and support the medical team could provide. Her temperature was closely regulated, nutritional physiologists administered essential vitamins and minerals through a tube, and oxygen was blown into her nose. But I knew that the hospital couldn't meet all of her needs. Infants' brains need love and nurturing if they are to develop properly. Food, warmth and clean air simply aren't enough. Nor was it enough for the physiologists to add my expressed breast milk to her feeding tube. She couldn't feel indirect love; she needed direct, skin-to-skin contact.

Even full-term babies don't have fully developed brains at birth. Rather, their brains continue to develop as they interact with other people. So insufficient interaction results in insufficient development. In the middle of the twentieth century, several studies found that infants and young children in hospitals and orphanages have a tendency to become passive, lose the ability to walk or talk and stop gaining weight. Some even die. While such children generally receive adequate food, clothing and warmth, they are deprived of love. Hence, a doctor named René Spitz concluded that all children need nurturing if they are to develop normally. It was later discovered that the brains of children who have been emotionally neglected are smaller than those of children who are raised by loving parents. Children learn when they are greeted with a smile the first time they try to walk a few steps or when they are comforted after a fall. This learning process involves the creation of hundreds of thousands of new neural connections, so the brain physically grows. On the other hand, neural connections can wither and die if a child fails to use them, especially in the first two years of life.

Further studies have shown that brain development is impaired even when children are raised in attentive, caring institutions. One or two permanent caregivers who are always there for infants and toddlers simply seem to be more conducive to full development than twenty or thirty employees who work in shifts, no matter how affectionate the latter may be. One study monitored the progress of two groups of orphaned children: those in the first group went into foster care while those in the second were raised in an orphanage. After a number of years, the former group scored higher on an IQ test than the latter.

However, the brain controls every aspect of who we are and how we live our lives, so a loving, nurturing environment is crucial for more than just our intellectual development. Children who receive unconditional love while growing up are also more likely to be socially adept and empathetic than those who are neglected. So, in the long run, we'd all benefit if they were given it.

CHAPTER 6

INTELLIGENCE

Moderately wise
each man should be,
never too wise;
people who don't know
too much have the
nicest lives.

In 1933, the author Aksel Sandemose proposed what he called the 'Law of Jante'. In essence, his suggestion was that individuals shouldn't be too full of themselves because the welfare of the group is more important than individual achievement. This idea is typically Scandinavian, and it dates back much further than the 1930s. Indeed, the epigraph above is taken from a Viking-era poem called the *Hávamál* that was first transcribed in the thirteenth century.

Nevertheless, Scandinavians also accept that everyone is different. We understand that each individual has his or her particular strengths and weaknesses. We know that some have a great sense of humour, while others have a terrific memory, are more musical, can pick up a new language in an instant or are more adept at sports. Yet, we are not so willing to accept natural differences when it comes to intelligence.

In this respect, we believe that everyone should be included and everyone should be equally impressive. But is there really a measurable factor that we can call 'intelligence' and, if there is, does it tell us anything about ourselves?

Intelligence can be defined in various ways, so there are many possible answers to this question. Moreover, some people believe that it is not a single, definable characteristic at all but should be divided into various constituent parts, such as social intelligence, language-related intelligence, musical intelligence and so on. Others argue that this undermines the whole concept of intelligence and insist that we should focus on the classic definition – an aptitude for abstract thought – and not include applied or social skills.

Some people who are deemed to be highly intelligent on the basis of this classic definition are also accomplished tennis players, while others trip over their feet every time they set foot on the court and never manage to hit the ball. Some of them have extraordinarily good memories while others can scarcely remember what they had for breakfast. All they have in common is that they seem to be extremely accomplished at acquiring knowledge, solving problems and thinking logically. No more, no less.

IQ

To be of any value whatsoever, a test of intelligence must measure a person's reasoning and aptitude for abstract thought independent of their ethnic and socioeconomic background, education and gender. The ideal would be a test that produced very similar results every time a certain individual took it, with no significant variations over time.

IQ stands for 'Intelligence Quotient', with 'quotient'

another term for 'ratio'. So the first IQ tests supposedly generated ratios of the participants' mental age to their chronological age (with the results multiplied by 100 for ease of comparison). They were not absolute measures of the participants' intelligence, per se. IQ is no longer calculated in this way, although the name lives on. Today, everyone who sits an IQ test is measured against a reference group where the average is set at 100. An individual's score is based on how they perform in relation to that reference group. Hence, if a whole population were to sit the same test, the results should generate an almost perfect bell curve. Approximately 50 per cent of the sample will fall between 90 and 110, 68 per cent will fall between 85 and 115, and about 96 per cent will fall between 70 and 130. The 2 per cent who record an IQ below 70 are classified as intellectually disabled. The 2 per cent who score more than 130 are eligible for membership of the international high-intelligence organisation, Mensa.

For almost a century, countless teams of researchers have striven to make IQ tests as good as they can be. Yet there are still dozens of rival systems, each with its own supporters. In other words, after a hundred years and millions of test papers, even the experts can't agree on the best way to test intelligence.

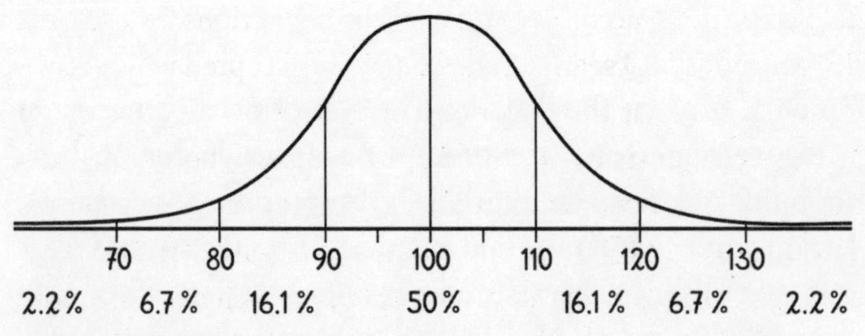

Figure 19: A typical IQ graph, with 50 per cent of the sample group falling between 90 and 110.

The most commonly used tests aren't designed to measure the participants' level of formal education or their knowledge of reading, writing and arithmetic, so the questions are often based on abstract pattern recognition. Nevertheless, participants from some cultures may never have used a pen before arriving at the testing centre, which puts them at an obvious disadvantage in comparison with those from Western nations. Moreover, as with any other exam, performance can vary depending on how the participant is feeling when they sit the test. A broken heart, financial worries, insufficient sleep or a lack of food might result in a loss of concentration and a score of 105 (average) rather than 115 (intelligent). Finally, the designers of IQ tests seem to assume that they will always be held under optimal conditions – in bright, well-ventilated, quiet exam halls. Of course, this is often not the case, especially in the developing world.

Understandably, then, critics of IQ testing insist that it is a far from perfect measure of intelligence. Nevertheless, many countries continue to use it in medical diagnostics. As mentioned earlier, anyone with an IQ below 70 is classified as having an intellectual disability, but there are subcategories all the way down to less than 20. For instance, in Norway at least, a person with an IQ below 55 cannot be held criminally accountable for his or her actions. Therefore, the concept of IQ seems to be generally accepted when dealing with those at the lower end of the scale.

The situation is rather different for those who score more than 130. Few people view high intelligence – at least as measured in an IQ test – as synonymous with wisdom. The latter is seen as a broader concept that includes common sense and acquired knowledge as well as abstract reasoning. It is linked to what has already been learned, whereas IQ is primarily related to the potential to learn. Even members of Mensa can find it difficult to realise that potential.

High IQ – so what?

In practical terms, is it really any use to know that the next figure in a sequence will be a parallelogram divided into four white and black segments? Does having the ability to work that out during an IQ test prove that a person has an excellent memory or knows how to be a good friend, parent or spouse? Of course it doesn't. You can easily find a homeless person with a high IQ, or a millionaire businesswoman with an average IQ. Nevertheless, in general, 'intelligent' people – that is, those with IQs above 110 – are often able to come up with solutions to problems that others fail to see. That alone means they are more likely to find a good job, earn a high salary, live in a comfortable house and even enjoy a harmonious family life than those who are lower down the scale.

One study found that 55 per cent of people with a low IQ had dropped out of high school, while all of those with a high IQ had completed their secondary education. That could explain why 30 per cent of those with a low IQ were in a bad financial situation, while only 2 per cent of those with a high IQ were experiencing similar trouble. Perhaps more surprising is the fact that IQ appears to correlate just as well with domestic stability. For instance, women with low IQs are four times as likely to have children out of wedlock as those with higher IQs, while mothers with low IQs are eight times as likely to receive benefits as their counterparts with high IQs. Similarly, individuals with below-average IQs are twice as likely to go through a divorce as those who are above average.

All of us have probably seen a handsome boy or a beautiful girl walking down the street and secretly hoped that they're not too bright, because it doesn't seem fair when someone 'has it all'. But nature isn't fair. Recent research has established a link between appearance and intelligence. Simply put, it seems that attractive people – as a group – are smarter than the rest of us. One study followed 17,000 British children between 1995 and 2011, during which time each and every one of them sat eleven intelligence tests. Meanwhile, multiple teachers objectively evaluated their appearance. Similarly, an American study followed more than 20,000 children over the course of eight years. Again, the students sat a number of intelligence tests and independent assessors evaluated their physical attractiveness. In both the British and the American study, there was a clear correlation between good looks and intelligence.

In the years since these reports were published, many researchers have attempted to explain their remarkable findings. Some have argued that intelligence and attractiveness are both manifestations of general good health, citing the old adage, 'Healthy mind, a healthy body'. Others believe that the correlation is linked to natural selection. Intelligent men with good jobs and secure financial situations tend to marry attractive women and vice versa. And because both intelligence and physical features are heritable, their children tend to be both attractive and smart.

Even if IQ tests only test our ability to solve seemingly meaningless conundrums – such as recognising the next geometric figure in a sequence – multiple studies have shown that the results can be used to predict a person's aptitude for tasks that demand high-level language and mathematical skills or a good memory. These clear correlations have led some researchers to argue that IQ tests actually measure a sort of general intelligence, known as the g-factor. When this is tested, subjects achieve the same results regardless of whether the test uses words, series of numbers or abstract figures, and regardless of whether it is taken orally or in writing, individually or in a group. A high g-factor score usually correlates to subsequent success in school or at work.

In conclusion, though, a high IQ doesn't equate to high intelligence; rather, it's just one measure of intelligence. Moreover, while intelligence might help you achieve your goals, many other factors are sure to come into play, too. Nobody's life is preordained by their position on an IQ scale.

Long-headed and short-headed

Today, most of us scoff at the nineteenth-century pseudo-science of phrenology, which attempted to estimate people's intelligence from the shape of their heads. However, a number of recent, scientifically rigorous studies have claimed that there is indeed a link between brain size and IQ. We know there are exceptions – including Einstein – but on average, it seems that extremely intelligent people generally have larger brains than those of more average intelligence (according to their relative positions on an IQ scale). That is, if you add together all of the brain sizes of a large group of intelligent people and compare that number with the aggregate brain sizes of an equal number of less intelligent people, the first figure will be significantly larger. And the difference between the two groups is even more pronounced when just the frontal lobe (which is important for logic and abstract thinking), the temporal lobe (which, among other things, is important for memory) and the cerebellum (which is most concerned with motor coordination but also has a role to play in thought processes) are measured. Interestingly, no study has managed to establish a link between the amount of white matter – which houses the brain's signal pathways – and IQ. By contrast, the correlation between the amount of grey matter – where the neurons themselves are found – and intelligence is undeniably strong.

This holds true for children as well as adults, with the correlation especially clear when only the cerebral cortex at the front of the frontal lobe is measured. Yet, this doesn't mean that we should start assessing everyone's intelligence on the basis of their brain scans alone. While the correlation between the amount of grey matter and intelligence now seems indisputable, at most it accounts for only about 20 per cent of the difference in intelligence between one person and another.

Researchers around the world are now exploring how the brains of highly intelligent people actually work. Over the last twenty years, a number of studies have found that intelligent people employ smaller amounts of their cerebral cortex when solving problems than those with lower intelligence. The general conclusion is that their neural activity is more focused.

It takes a huge amount of effort to reach the top in any field: you don't get to be a four-time Olympic champion like Mo Farah by lounging around on the sofa. But it's certainly an advantage to be born with a good pair of lungs and a strong competitive instinct. Obviously, very few people would be able to match his achievements, even if they trained just as hard and in precisely the same way. That's how it is with the brain, too. We're all born with different potential, so it's simply a case of realising as much of it as you can.

Nature or nurture?

Most of the variation in intelligence between individuals in a particular population seems to be due to genetic inheritance rather than environment. For instance, the circumstances in which adopted children are raised don't have a noticeable effect on their IQ, as long as they receive good care. Instead, as they grow up, their IQs tend to approximate those of their biological parents, even if they've never met them. Therefore, financial or social context seems to have no lasting impact on intelligence, at least as measured by IQ tests in the Western world. A number of studies have found that it has *some* effect on children's IQs, but this seems to disappear as they reach adulthood.

Nevertheless, the debate about the relative importance of nature versus nurture in determining intelligence shows no sign of abating. In part, the controversy has been fuelled by the fact that average IQ has increased over time. Now, this is a complicated subject because the average IQ of a group of people who sit the same test must be 100 on account of the way in which IQ tests are scored: individual participants are always marked in relation to the group's average, so half of them will always be in the central, average band, while smaller proportions will be in the low and high sectors. That said, the setters have made IQ tests steadily harder over the years, and someone who scores 100 on one of today's tests would almost certainly score higher if they were to sit one from the 1940s.

This is simultaneously both fascinating and frustrating, because we're not entirely sure why the human population has become more intelligent (or at least more adept at answering questions during IQ tests). The general consensus is that it's mostly due to natural selection. Throughout much of the previous century, the children of wealthy, healthy parents were more likely to survive into adulthood than those of impoverished parents, even in the West. Therefore, as long as we accept that affluence is linked to intelligence, the genes of intelligent people were more likely to be passed down through the generations, leading to a gradual increase in the average intelligence of the population. However, some researchers are now suggesting that this process has already ceased and may soon go into reverse. Whereas the most affluent sectors of society used to produce the most descendants, the opposite is now the case, partly because they start trying for children much later in life than the less well-off. Moreover, in the West at least, medical advances and the development of the welfare state mean that children from poor backgrounds now have a much better chance of surviving into adulthood and passing on their genes.

Therefore, it's safe to say that environmental factors – as well as genetics – have played a role in the general increase in human intelligence, and not just in terms of giving the poorer members of society more opportunities to reproduce. Over the last century, there's been a clear increase in average human height due to better living conditions and better nutrition, both of which have also been linked to improvements in brain function. Similarly, our daily routines are very different from those of our great-grandparents. We undertake far more tasks that demand abstract thought and reasoning than they did, and we are less reliant on practical, hands-on skills. Whereas they used to scrub their laundry with a brush and feed it carefully through a mangle, we have to decode complex symbols on the washing machine and our clothes in order to select the best program. While most of us would lose a finger if we tried to sharpen a knife on a grinding stone, they would be left baffled by the buttons on a TV remote control. All of these are environmental factors, and they have probably contributed to generational improvements in the type of thinking that IQ tests measure.

Nevertheless, some individuals are sure to be disappointed when they receive the results of an IQ test. So, is there anything they can do to achieve a better score next time? The honest answer is 'not much'. In general, an individual's IQ will remain unchanged once they've reached adulthood, regardless of the career path they take or whether they become rich or poor. If there's a way to increase the g-factor, no one's found it yet.

That might sound depressing. After all, we've seen that, on average, people with high IQs are more likely to succeed in life than those of lower intelligence. However, you can still have a very big impact on your prospects if you're prepared to put in the necessary effort. Studies have shown that Chinese-Americans, Japanese-Americans and American Jews all overachieve when compared to white Americans. For instance, an averagely intelligent Chinese-American – with an IQ of 100 – is likely to achieve just as much in life as a white American with an IQ of 120. In other words, the effort we put into reaching our full potential can be just as important as the potential itself.

In light of this, some psychologists have divided intelligence into two distinct components. The first of these is known as 'fluid intelligence' and it is the form that we have focused on throughout this chapter. It remains stable throughout adult life – at least until the onset of dementia or some other form of brain injury – and relates to how well the brain functions from a biological perspective. The other form – which is known as 'crystallised intelligence' – relates to your ability to make the most of your opportunities and utilise the skills and knowledge you have learned over the years. Therefore, although there's nothing you can do about your fluid intelligence, you can increase your crystallised intelligence every day and make the most of your potential.

The downside of high intelligence

Schools are designed for the average student. Those with very low and very high intelligence tend to be neglected. Consequently, high intelligence does not guarantee smooth progression through the education system. Indeed, a child with extremely high intelligence may require more attention than an average student. For instance, they may become bored and restless if they understand an assignment immediately and have to wait for the rest of the class to catch up over the course of the next few days. Over the long term, slow classroom progress and too few intellectual challenges can cause such children to develop poor work habits, which in turn can result in them failing to reach their full potential.

In addition, the most intelligent children often have trouble fitting in socially. If we think in terms of how IQ was originally calculated – as the ratio of mental age to chronological age – then it's hardly surprising that an eight-year-old child with a mental age of thirteen doesn't play so well with her classmates. Indeed, by that age, she might have lost all interest in play. Maybe the extract from the *Hávamál* that opens this chapter has a point. Maybe people who don't know too much have the nicest lives.

So, to be fair to child geniuses, should we create elite groups in elementary school in much the same way as we cater to the special needs of children with low IQs? Or would that merely serve to widen the gap between them and the rest? These are difficult questions, and there are no simple answers.

Artificial intelligence

If you score well in an IQ test, there's a high probability that your whole brain is functioning well. Therefore, such tests are useful tools for measuring human intelligence, as long as we define 'intelligence' as having an aptitude for problem-solving and a high capacity for logical and abstract thinking.

Machines are different. I'm sure a computer could be programmed to answer all of the questions that appear on a standard IQ test in the blink of an eye. Maybe such a program already exists. So, if we were to measure artificial intelligence in the same way as we measure human intelligence, we'd have to concede that even the humblest laptop is a genius. However, there are many other aspects of human intelligence. Consequently, even proponents of artificial intelligence have to acknowledge that no computer can be considered truly intelligent until it manages to perform *all* of the tasks that an average human cerebral cortex tackles every day.

At present, computers do just fine with well-delineated tasks like playing a game of chess or manoeuvring a trolley around hospital corridors. And the day is surely coming when a robot will be able to detect tears rolling down the cheek of a patient and recite a few pre-programmed words of comfort. But could such a response really be termed empathy if it's not generated by compassion? Computer engineers are a very long way from making artificial brains that work in the same way as our own, not least because we still know so little about how the human brain actually functions. Moreover, our brains have evolved over millions of years, so it's unlikely that they'll be replicated in a few decades of software research.

Chapter 7

CULTURE © THE BRAIN

Why did Stone Age humans carve petroglyphs? When I walk around Ekeberg, in Oslo, and look at the preserved rock carvings from 4000–5000 years ago, I am left in awe of the amazing human brain. The people who carved them lived in caves and tents made of animal hides, had a life expectancy of little more than thirty years and had to find their own food each and every day to avoid starvation. So why did they devote so much time and effort to the laborious task of carving figures into rocks? Why does the human brain value creativity, interpretation and imagination so highly?

Some people believe that culture must have emerged along with language and the ability to plan – that is, about 200,000 years ago, when *Homo sapiens* first walked the earth. However, the earliest tangible evidence of human culture is only 40,000 years old, which was around the time when we started supplementing our traditional tools – such as hand axes and picks – with fish hooks. It requires a certain amount of abstract reasoning to devise something as sophisticated as a fish hook, so humans from this era must have been thinking in much the same way as we do today. This conclusion is reinforced by the fact that they started painting the walls of their caves at precisely the same time.

Their stick figures of fellow humans, animals and boats are now considered fundamental aspects of our cultural heritage, even though they don't come close to rivalling the artistic achievement of the ceiling of the Sistine Chapel. Similarly, musical culture encompasses everything from *Don Giovanni* to bawdy drinking songs. Indeed, almost anything can be considered part of human culture, from our languages, manners, customs and traditions to our rules, regulations and morals, to politics, religion and sports. Society's elders teach us about these things, which means they're passed down from generation to generation. But society consists of a variety of groups, so there's a variety of cultures. Norwegians aren't born with skis on their feet. I had to be taught about my culture.

Together we're strong

We often think that human capacity is limited to what each individual brain can achieve, but many heads are usually better than one. Our brains not only enable us to make tools so that we can cultivate the soil more efficiently but allow us to communicate and teach others what we know. Once someone invents the wheel, the next generation doesn't need to reinvent it. Instead, they can work on improving the design. Then a later generation can attach the wheels to a cart. Then bicycles, trains and cars can be developed.

Many other species also use tools, but they rarely refine them over generations. That's because no other species has our capacity for cooperation and empathy. We have so-called 'mirror neurons' in our highly developed cerebral cortex that help us see ourselves in others. Those that burst into life when I scratch my chin are also active when I see you scratching your chin. We don't even need to do it at the same time. Several studies have suggested that these neurons also play a role in social understanding, and possibly even in empathy.

Human interaction and cooperation are also facilitated by our unique ability to talk, read and write. Our highly developed thought processes and language skills mean we are no longer slaves to our instincts. They enable us to ask questions, judge and adjust the way we behave towards ourselves and others, formulate rules, and ultimately organise ourselves into civilised societies. The ways in which we interpret, think and speak today are the culmination of generations of social rules, norms and values that have underpinned a series of cultures throughout human history.

Social networks

So, there would be no culture without our complex brain; but, in return, culture gives that complex brain the perfect environment in which to grow. It ensures the safety and security that our brains need if they are to continue developing over the course of the lengthy human childhood (which, in total, comprises almost a quarter of the average person's lifespan). Genes provide the foundation of the brain's structure and the basic functions that it is able to perform at birth, but immediately thereafter the newborn child's surroundings start bombarding its senses with new information that affects and shapes its young brain. Neurons send this information to the regions of the brain that are best equipped to interpret it, following paths through the rudimentary communications network that the genes have constructed. However, as the data continues to pour in, it creates denser, more numerous and more complex neural connections. When we're born, each neuron has about 2500 points of contact in the form of synapses. By the time we're two or three years old, each neuron has about 15,000 synapses! And all of those new points of contact develop in response to the sensory information our brains receive, and therefore our environment.

It's easy to see that the brain of a newborn baby – who can't even make eye contact – is not fully developed. Yet, in the first year alone, infants start to react to facial expressions and tone of voice by smiling at happy faces and crying in response to stern rebukes. Not long after, they learn to speak and think for themselves, but always within the context of the norms and rules that dictate what is right and wrong in their environment. Hence, the outside affects the inside, and the inside affects the outside.

It's precisely because so much of our brain develops after birth that nature's genetic grip over humans is looser than it is with any other animal. We have the ability to re-examine our genetically instilled instincts on the basis of what we acquire through socialisation, which has resulted in the huge range of diverse cultures that coexist in the human population today.

The moment when children start to understand that some people have utterly different thoughts and attitudes from their own is considered an important milestone in the maturation of the human mind. In general, this process begins at around the age of three or four, but I'm convinced that some full-grown adults haven't gone through it yet. Indeed, if everyone throughout history had developed a full understanding and appreciation of alternative cultures while still a toddler, maybe the United States would have a chief instead of a president, and Australians would be throwing boomerangs instead of cricket balls. The socialisation process means that we see 'our people' as normal and civilised, and others as strange, alien, maybe even uncultured. For instance, in some places, girls are taught it's unseemly to show their hair; in others, it's considered improper to hide it. Gradually, though, now that we are more familiar with a wide variety of different cultures than at any time in human history, we are learning to accept and respect them. Of course, we still have a long way to go, but we are starting to realise that we live in a complex, densely populated social world that will surely collapse without cooperation, negotiation and tolerance.

The social code

Our cultural norms rein us in and control us. They lubricate the social machinery. The rules relating to appropriate and inappropriate behaviour have been imprinted on every one of us as we've grown up. We use them as a template for our social conduct as we proceed through life, but we also make our own, personal rules with the help of the foremost part of the frontal lobe – the prefrontal cortex. The maturation of this part of the brain relies on it receiving the optimal amount of dopamine, the reward neurotransmitter – neither too much nor too little. If the level of dopamine starts to vary, a person may become impulsive or distracted.

Similarly, people with a damaged prefrontal cortex lose the ability to follow society's rules. They often become uninhibited, entirely governed by their own urges. If they want to pinch someone on the bum, they do so. If they think an apple looks good on the supermarket shelf, they grab it and start eating it; they don't bother to take it to the checkout. They do whatever occurs to them, whenever it occurs to them, regardless of how inappropriate their behaviour might be. Several studies have shown that individuals who don't have a fully developed prefrontal cortex can develop antisocial personality disorder and might even start to commit serious criminal offences. This is becoming a major concern for the criminal justice system. If someone's criminality can be attributed to a malfunctioning or undeveloped prefrontal cortex, is it right to punish them? After all, if someone doesn't understand the rules of the game, if they can't tell the difference between what's right and wrong, should society hold them to account for their actions?

Humans have learned to cooperate with each other to secure food, care for our children and protect ourselves, among many other things. This cooperation would be impossible without language. We are unique in the animal kingdom in that we have made communication easier for ourselves by using symbols. A semicircle with a straight line down the left-hand side represents a lower-case 'b'. Get rid of the top of the line and most of the semicircle and you're left with an 'r'. Join a couple of curved lines together and you make an 'a'. Put a dot above a short line and you have an 'i'. Replace the dot with a line that curves to the right then goes straight down and you've created an 'n'. Before you know it, you've written the word 'brain'. A few straight lines, curves and dots allow us to communicate our thoughts and feelings in dozens of different languages. When musicians decode other lines and dots, they are able to play precisely what a composer intends them to play.

The creative brain

One way in which we enrich our everyday lives is by telling stories. Our brain gives us the capacity to create, recite and understand all sorts of tales, which in turn aids the brain's development. The psychologist Donald Hebb discovered that rats who were raised as pets were better at solving problems than those that grew up in a cage. Other researchers have continued to explore this phenomenon and have proved that the brain benefits from a stimulating external environment. Given this research, it's reprehensible that ever more primary schools are being built like barracks around a small patch of tarmac with no consideration for architecture that would help to optimise the children's brain development. If laboratory mice and rats are provided with something as simple as a bundle of sticks in their sawdust, and especially if they are given a running wheel or a ladder to a second floor, they start to form more synapses between the neurons in their brains and their cerebral cortices thicken. Moreover, it's now believed that they also create new neurons. So just imagine what a more stimulating environment would do for human children. Similarly, there are many indications that the external stimuli which our culture offers in the form of books, music, theatre shows, architecture and interactions with other people can delay the onset of dementia, simply because it gives the potential sufferer greater intellectual reserves.

'Chop? That's a great name! That's going to be your name. You're going to be my friend, because there's no one else here. You and me, we're friends.' In this extract from one of Anne-Cath. Vestly's children's books, a young boy is talking to a plant that looks like a miniature person. No other species possesses the amount of creativity that's necessary to think in this way. Chimpanzees don't find sticks on the forest floor and befriend them, and dolphins don't chat away to dolphin-shaped lumps of coral. Part of what makes us human is that we have the imagination to do these things.

Whenever you come up with something new, you're exercising your creativity, which demands a certain degree of intelligence, critical thinking and selectivity. However, you don't need a high IQ to make your mark in a creative field. Andy Warhol – whom many people would term an artistic genius – had an IQ of just 86. As was discussed earlier, our brain filters out much of the constant stream of information from our senses before it reaches our consciousness. This enables us to focus on the most important tasks – an important aspect of everyday life. However, creativity rests on opening ourselves up to sensory information and memories that seem to have no immediate value or use. This process helps us make connections between things which, at first sight, seemingly have nothing to do with each other.

Modern technology such as magnetic resonance imaging (MRI) and positron emission tomography (PET) allows us to witness the brain's creativity in action. The former lets us see which regions of the brain receive the most blood supply during problem-solving, while the latter shows us which parts use the most sugar. Specific regions are highly active when a subject is set tasks that involve motor skills, skin sensation or language. By contrast, when creativity is tested, many sections of the cerebral cortex are involved. This is understandable, because creativity requires cooperation between many different functions that are located in various parts of the brain, including in one or other of the hemispheres.

For many years, scientists have insisted that creativity is located primarily in the right half of the brain. However, there's little hard evidence for this. While certain parts of the right-hand prefrontal cortex seem to play a more active role in creativity than their counterparts on the left, this may be due to the simple fact that much of the left side is devoted to language. In any case, the rest of the frontal lobe on the left seems to be just as active as the right during creative tasks, and the same is true of the parietal lobes. In other words, creativity resides in both halves of the brain.

Does Mozart make you smart?

Many studies have explored the effects of music on the brain. For instance, does listening to Mozart make people more intelligent? In one study, the performance of a group of students who were set problems that required good spatial perception improved for the first fifteen minutes after listening to Mozart, which led to a rush for his CDs after newspapers reported the results. Pregnant women started playing Mozart to their bellies in the hope that doing so would produce smart babies. Indeed, the Governor of Georgia went so far as to ensure that every newborn in the state received his or her personal CD of classical music. Other studies suggested that rats' ability to make their way through mazes increases if they are played Mozart as foetuses. Some water treatment plants even started to pipe Mozart through their facilities because of claims that this stimulates the bacteria to break down the waste at a faster rate.

However, despite their best efforts, several groups of researchers have been unable to substantiate the claims of the first study. Sceptics now claim that the so-called 'Mozart effect' is actually just a clever marketing ploy to boost the sales of CDs, teaching materials and books that champion the alleged intelligence-promoting properties of his music. One particular piece – the Sonata for Two Pianos (K448) – has received special attention. Those who have faith in the Mozart effect claim that this piece harmonises especially well with the body's natural rhythms, including brainwaves and heartbeat. Moreover, several small studies have suggested that it should be played to patients with a specific type of epilepsy that cannot be treated with conventional medication.

Although further investigation is needed, at least listening to Mozart doesn't have any known negative side effects. And while we cannot conclude that listening to his sonatas is sure to make you smarter, musical education does seem to have a role to play in increasing young children's general intelligence. This should come as no surprise, since *all* learning has a positive impact on intelligence. It probably doesn't make much difference whether they try to remember all the words to the latest Taylor Swift album or learn how to play 'Twinkle, Twinkle, Little Star' on the recorder.

Preference for a particular type of music is a significant component of individual identity. What sort of people listen to classical music? Actually, the question should be 'What sort of people *claim* to listen to classical music?' because scientists have no option but to trust that the subjects they test in their studies are telling the truth. Anyway, people who say they listen to classical music tend to have better qualifications and drink more wine than, for example, hip-hop fans, according to British research. But does that mean that listening to classical music guarantees high educational achievement, or is this a case of people with a high level of education simply adopting the habits of their immediate social circle? We know that the brain is susceptible to influence, and of course one of those influences is music. How can a single piece of music move one person to tears of joy and leave another hearing nothing but discordant noise? Many unanswered questions remain, but musicians and brain researchers are starting to collaborate on a number of exciting projects in a bid to find the answers.

In the meantime, we already know that the brain does not

interpret singing and speech in the same way. This means it's possible to lose the ability to speak – following a stroke, for example – yet retain the ability to sing. People used to believe that this was because the music zone was located in the right half of the brain, while the language zone was in the left half. Now, though, we know that, in general, the left half looks after the lyrics and the rhythm when we sing, while the right half deals with the melody. Sound waves that reach our eardrums are interpreted first in the auditory cortices in the temporal lobes, then further interpretation follows in other regions of both hemispheres. These regions work as a team to help us recognise and understand what we have heard. Meanwhile, the limbic system links emotions to the sounds, which is obviously important for whether we enjoy listening to them or not.

Therefore, music affects how we feel. Our musical preferences aren't set in stone – they depend on what we're doing and what kind of mood we're in at any given time. Regardless of whether you're listening to Lady Gaga or Mozart, it has an effect on the human brain that we do not find in other animals. Whenever you listen to music, the portion of the basal nuclei known as the nucleus accumbens – the brain's centre of love and desire – is activated, which triggers the release of dopamine from a group of neurons in the brainstem. This signal pathway is called the reward pathway and it's activated in a variety of situations. For example, dopamine is released when a chocoholic eats chocolate, when an addict injects heroin and when you see that someone has liked your latest upload on Instagram, although the amount that's released and hence the degree of happiness you feel depend on how surprised you are. Consequently, if you stumble across a new song that you immediately love, your brain releases more dopamine than it does when you listen to a favourite song that you've heard a hundred times before.

Several studies have shown that we perform better at boring, repetitive tasks – and finish them faster – when we listen to *any* sort of music, regardless of personal preference. But you should hit the pause button before trying to learn something new. Mentally demanding tasks require focus, so if you're studying a new language or trying to solve a difficult Sudoku puzzle, it's probably best to turn off the music. However, a group of nurses told me that a former colleague used to listen to music in the operating theatre while performing brain surgery. If you're good at what you do, then it's fine to have music on in the background, even during a demanding task. Indeed, studies in the *Journal of the American Medical Association* and other esteemed journals found that surgeons who listened to their favourite music in the operating theatre worked both faster and more accurately than when they operated in complete silence. Similarly, a little background music seems to boost creativity.

This may be linked to the fact that tension is relieved when we listen to favourite pieces of music, again irrespective of the type of music the listener prefers. For instance, my little sister has heavy metal blasting out of the earphones while she studies. Each to her own. But if you need to concentrate, you should always choose music you know well. As mentioned above, hearing new music for the first time prompts a massive release of dopamine, which draws most of your attention to the music rather than the task at hand. So try to stick to your favourite playlist whenever you're working or studying. If you just can't wait to hear something new, choose an instrumental or something with very few words as these are less distracting than songs with lots of lyrics.

The same almighty God

Almost every culture around the world has developed some form of religion. Some people argue that societies which practise religion have an evolutionary advantage because they rein in egotistical, antisocial behaviour by promoting belief in eternal, watchful ancestors and vengeful gods. In this explanatory framework, the gods take the form of powerful, overprotective parents who see and hear everything we do and think. People fear them and obey society's rules to avoid their censure.

In the period when the ancient Greeks were establishing the world's first democracy, the ancestors of the Vikings were holding big parties during which they sacrificed animals and poured the blood over themselves. How can two groups with identical brains create such contrasting cultures? Well, the fact of the matter is that they were not so different after all. The Greeks also sacrificed animals; and while the Vikings drank copious amounts of beer, the Greeks probably matched their alcohol intake through their consumption of wine. Moreover, both of their pantheons of gods – and many others besides – were associated with important events or phenomena that humans didn't understand. Take thunderstorms, for instance. The Norsemen believed that thunder was heard whenever Thor wielded his hammer while driving across the sky in his goat-drawn chariot. Greek children were taught that the Cyclops fashioned thunderbolts for Zeus to use as weapons in his war with the Titans. The Romans believed that Jupiter was throwing lightning bolts at his enemies every time there was a thunderstorm. In Hindu mythology, the storm god Indra summons lightning with a club as he is pulled across the sky in a golden chariot. While his chariot may be more ornate than Thor's, there's no denying that the human brain managed to create astonishingly similar myths in very dissimilar environments at almost exactly the same time.

Eventually, as belief in the old mythologies started to fade, new gods emerged, although some religious historians claim that the God of Judaism, Christianity and Islam was originally just one member of an earlier pantheon – a volcano god who controlled thunder and lightning. Either way, most people once again conformed to a single pattern, this time by placing their faith in one almighty God.

Different cultures, similar stories

Similar patterns also emerge in the world's myths and fairy tales. For instance, the Brothers Grimm's Cinderella bears a remarkable resemblance to the Norwegian Katie Woodencloak, who also entrances a prince to escape from a life of gruelling domestic servitude. Our heroine is helped by a dove in the German version, a bull in the Norwegian version ... and a fairy godmother in Charles Perrault's French version, which Walt Disney preferred when he decided to turn the story into an animated feature film. In each of these versions, Cinders/Katie has improbably petite feet, so when she loses a shoe after meeting the prince, he knows he is on fairly safe ground when he promises to marry whoever it fits. This is a legacy of the Chinese version of the story – Yè Xiàn. Tiny feet were so revered in medieval Chinese culture that girls were forced to suffer the agony of foot-binding, so the paragon of beauty Yè Xiàn had to have the smallest feet in the land. Of course, Cinders/Katie/Yè Xiàn always marries her prince and lives happily ever after at the end of the story.

Understanding the abstract

Can geometrical figures be works of art? In *Guernica,* Pablo Picasso managed to evoke the horrors of the Spanish Civil War in a series of triangles, semicircles and jagged lines. He represented a human face with an asymmetrical triangle, a curlicue (for an ear) and two arcs (for eyes). If I saw such a creature in reality, I'd be really scared! But when I look at *Guernica,* I see a fellow human in anguish. Our ability to interpret and understand abstract art, musical compositions and installations in this way reveals the immense complexity of the human brain and the power of its reasoning.

Crazy or brilliant?

Our brain is so complex that misconnections can occur, however. We're already familiar with this problem from our dealings with modern gadgets: the more complicated they are, the more likely they are to malfunction. In humans, this complexity is most apparent in highly creative artists, who are often condemned as crazy and praised as brilliant in equal measure.

Under normal circumstances, the human brain – specifically the cerebral cortex and the thalamus, which sits at the top of the brainstem (see Fig. 1) – sees to it that we receive essential information about what's going on around us in manageable portions by filtering out all of the extraneous detail. For instance, this filtering process allows us to understand the essence of a message without any need to analyse each word individually. It also lets us conduct a conversation in a shopping mall while our ears are bombarded by twenty or thirty other conversations, a cacophony of songs blaring from every store, the rattle of escalators, babies crying and countless other sounds.

However, it should be remembered that the thalamus is not the most sophisticated part of the human brain. In most of us, it tackles the problem of sensory overload as it always has – by blocking the vast majority of the information we receive – which is one reason why so many people share a similar world view. Recently, though, a team of Swedish researchers found that both highly creative and schizophrenic people have far fewer receptors for dopamine in the thalamus than the rest of us. Hence, their filters are relatively inefficient, so they are able to draw on information and feel emotions that the rest of us never experience, which may explain their high levels of creativity as well as their atypical view of the world. Exploring the link between creativity and mental illness in this way is a new, exciting field of research. The sample group in the Swedish project was too small for the team to draw any firm conclusions, but its findings are already shedding light on why some people are creative geniuses, why others hear non-existent voices or see hallucinations, and why some have a foot in both camps.

Vincent van Gogh painted some of his most audacious works during his confinement in a psychiatric hospital, while Edvard Munch acknowledged that his poor health and nervous disposition were prerequisites for his art. *The Starry Night* would have looked very different had the former not suffered a nervous breakdown, and the latter never would have painted *The Scream* had he been calm and content rather than beset by anxiety at the time.

Chapter 8

Eating with Your Brain

We've all grown up knowing that we have taste buds on our tongues that detect sweet, salty, sour, bitter and savoury (otherwise known as umami) flavours. And it's now thought that our intestines are also able to detect sweet substances, while some studies have suggested that much of what we call our sense of taste is actually located in the palate rather than the taste buds. But it should be remembered that we wouldn't be able to smell or taste anything without the brain. Taste buds, no matter where they're located – on the tongue, in the palate or even in the intestines – do not provide us with a taste experience on their own. Both taste and smell become meaningful only when the brain has interpreted the sensory information it receives. Only then do we taste in the true sense of the word. All of the choices you make about what you should put in your mouth are made by your brain. You eat with your brain.

Ancestral eating habits

But if that's the case, why don't we eat better, healthier food? Why is every trip to the supermarket a battle to resist the temptation to buy junk food and sugary treats? The answer is simple: the older and more primitive parts of your brain make you crave sweet and salty foods. Moreover, they conjure up excuses for why you are entitled to a little treat every once in a while. So, you can blame your ancestors the next time you feel yourself drawn to the rack of sweets at the checkout. Evolutionarily speaking, it was beneficial for us to crave salty foods as they provided essential minerals, umami flavours to ensure we got enough protein in the form of meat, and sweet and fatty foods, which provided an immediate energy boost and helped us build up reserves to see us through leaner times. The ability to store fat was a distinct advantage for our ancestors, not a health risk. When the next meal depended on a successful hunt rather than the corner shop or the local takeaway, they had to be prepared for an empty table – maybe for weeks on end.

Nevertheless, our highly developed cerebral cortex – and especially our prefrontal cortex – allows us to resist these primitive urges. For instance, we can use our memory to remind ourselves that chocolate and potato chips aren't healthy. Learning is key to banishing our ancient cravings for sweets and fat.

Food and sex

The Norwegian neurologist Are Brean begins many of his lectures by pointing out that we all depend on two liminal activities – that is, activities during which something penetrates the human body's boundaries. Both of them are essential to our survival: eating ensures that the individual survives; and sex ensures that the species survives. Allowing something to enter the body is always potentially risky, but our brain can draw on several million years' experience to keep us safe. It works hard to ensure that whatever we put in our mouth is not poisonous, and also tries to guide us towards food with some sort of nutritional value.

Smell plays a crucial role in this gatekeeping process. Our sense of smell is often unfairly criticised. Although dogs have twice as many olfactory genes as we do – and therefore much more sensitive noses – our brain is still able to use the information our nose provides to avoid food that might hurt us. Indeed, in many ways, our sense of smell is more sophisticated than any dog's, because our brain has much more capacity to *interpret* all of the olfactory information it receives. So, while dogs can smell little more than 'food', 'potential sexual partner' and 'competitor who has strayed into my territory', we can smell 'Christmas', 'summer holidays' and 'spring planting'.

Moreover, our sense of smell has an ally in its constant mission to protect us from poisoning. It's not only the smell of mould that sets off alarm bells in most people, but the *sight* of the blue-green fungus. Yet our brain is so highly developed that it doesn't just impose a blanket ban on anything with that colour and odour combination. At some point in human history, someone must have eaten some mouldy cheese – probably by accident – and discovered that they suffered no ill-effects. In fact, they found it delicious. Once that barrier was crossed, the rest of us could learn from their example and overcome our inherent aversion to mould – at least in a slice of Stilton or Roquefort. The same is true of fermented foods. Just because most Norwegians enjoy some traditional salted, fermented trout every now and then, that doesn't mean they're going to wolf down rotten bread and apples. It just means that they've learned that one particular dish, when it's prepared under controlled, hygienic conditions, won't do them any harm ... despite its rancid smell!

The joy of food

In an evolutionary sense, the eating habits that are causing so many health problems today helped our brains to grow larger and ever more complex. So, to some extent, they're the reason why we've been able to become the dominant species on the planet. Our early hominid ancestors ate foods with low energy densities, such as root vegetables, leaves and fruit. If their brains had been as large as ours, they would have had to eat all day long. Then along came *Homo habilis,* who mastered the use of fire and thus could eat meat without the fear of dying from infections. Plus, heating food dramatically increases the amount of energy that can be obtained from it. This allowed *Homo habilis* to meet all of their energy needs with fewer meals, which in turn meant that all of their thinking didn't need to focus on where they might find their next dinner. Since then, all the way up to our own species, *Homo sapiens,* the hominid brain has continued to grow primarily because of our increasing consumption of ever more energy-rich food. Unsurprisingly then, as far as it's concerned, the more calories, the better.

In other words, the human brain is constantly hungry. That's the price we pay for being the world's smartest species. Pound for pound, the brain needs more energy than any other organ, so it bathes itself in dopamine whenever we eat – or even see – sugary or fatty foods. This happens because the primitive parts of the brain are under the misapprehension that these foodstuffs are still in short supply, as they were 200,000 years ago. In that sense, it hasn't been keeping up with the times. Our brain is a product of evolution, and evolution is a slow process – far slower than advances in agriculture, which have made ever more energy-rich food available to ever more people over the last 10,000 years. So it continues to reward what is now an unhealthy lifestyle, even though this harms rather than benefits modern humans. Fortunately, the more sophisticated parts of our brain understand this and have the capacity to do something about it. We know what's healthy and what's unhealthy. So, even if your primitive reward centre urges you to keep gobbling down sweets, salt and chips, you *can* resist the pressure, because the more sophisticated parts of your brain are able to override the dopamine effect. If this weren't the case, every last one of us would be an obese slave to the food industry.

The foods that the primitive parts of the brain crave can ruin not just our teeth and our waistlines but the brain itself. Fat is stored as plaque inside the arteries throughout the body, including those within or leading to the brain. If a lump of plaque comes loose and blocks a cranial artery – or if the build-up of plaque is so extensive that no blood can get through – you have a stroke. Ultimately, this can lead to vascular dementia.

Addicted to sugar

So, in the short term, the brain rejoices whenever we consume a little sugar, salt or fat. Over the long term, however, we need to eat ever more of these foodstuffs to feel the same effect. Ironically, then, people who routinely try to increase their pleasure by eating lots of cake tend to get very little enjoyment out of doing so because they become desensitised to the constant release of dopamine. Indeed, rather than eating the cake to feel happy, they have to eat more and more of it just to stop feeling sad. On the other hand, if you can manage to resist the craving and restrict yourself to one small slice a week, you will experience unmitigated joy the moment the cake touches your lips and dopamine is released in your nucleus accumbens (see Fig. 20).

Marketing experts know their neuroscience

The food industry is well aware that we crave fat, salt and sugar. Over the years, it has crammed ever more of these substances into its products in a deliberate attempt to send the brain's reward system into a frenzy and demand more.

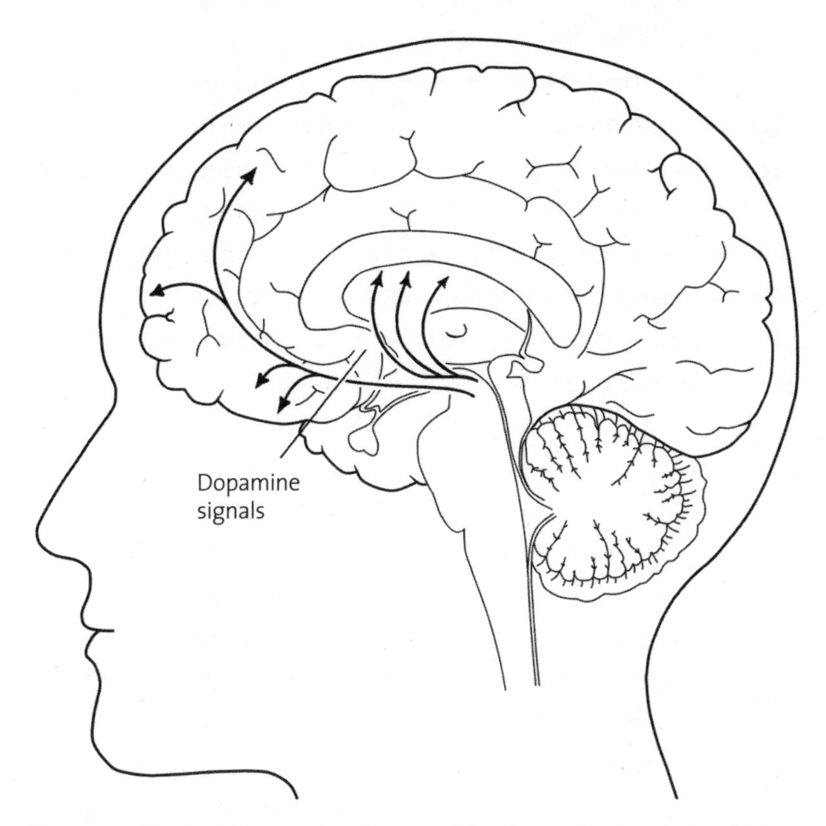

Figure 20: The brain's reward system consists of neural networks in which dopamine acts as a neurotransmitter. The dopamine signals radiate out from the midbrain to the basal nuclei, the limbic system and the cerebral cortex. Those that reach the limbic system travel via the nucleus accumbens, the brain's principal centre for love, rewards and desire.

And it is helped by the fact that not all of this craving is an automatic hormonal response in the most primitive parts of the brain. Some of it is learned. Many parents not only allow but actively encourage their children to reward themselves with unhealthy food. For example, primary-school children are often told they can have a chocolate biscuit as soon as they finish their homework. Of course, once this system of a sweet reward for virtuous behaviour is firmly established, it is very difficult to break, even in adulthood.

On the other hand, while the food industry is adept at exploiting the human brain's inherent susceptibility to temptation, we are astoundingly well equipped to resist. In addition to ensuring the success of our species by guiding us towards high-energy food, our brain has helped us find a varied diet. It tells us when we're full and even when we've had too much of a particular ingredient. The food industry – which never wants us to realise when we've had too much of anything – has researched this extensively and come up with a partial solution. It has learned that our brains cannot tolerate too much of any strong flavour, so its products tend towards the bland. Like most Norwegians, I would have no hesitation in saying that my mum's moose steak with homemade gravy tastes infinitely better than any hamburger. Yet, while I could easily eat a quarter-pounder, I would never be able to consume that much moose meat in one sitting – the flavour is just too rich. By contrast, my brain doesn't emit a 'stop' signal or tell me I'm sated when I'm halfway through a burger because there's scarcely any flavour while I'm chewing and absolutely no aftertaste. While our brain encourages us to eat small portions of a wide variety of ingredients, the food industry counteracts this by serving up slabs of largely tasteless fodder. However, simply being aware of this strategy can help us make more carefully considered food choices.

At least a hamburger eventually fills you up, whereas we seem to be able to consume other modern foodstuffs, such as potato chips or ice cream, in almost limitless quantities. That's because the brain reacts to a host of other factors in addition to total calories when assessing whether we've had enough or should just keep munching away. For instance, if you eat something that melts rapidly on your tongue, your brain is fooled into thinking that you've eaten less than you actually have. Similarly, high-sugar fizzy drinks are dangerous because the brain is less proficient at totting up the calories when they're consumed in liquid form.

The end result is that we consume far more calories than we need. And this problem is exacerbated when we start to look at food combinations. For instance, drinking alcohol increases the chance that you will opt for fatty foods, and eating fatty foods increases the chance that you will want a beer rather than water with your meal. At least, that's how it works with rats.

Advertising

If you don't want to be a passive consumer who just pigs out on whatever those wily marketing experts have decided to promote this week, you should learn a bit more about your brain. After all, as we have seen, the best marketeers are already one step ahead of you.

Robert Woodruff, president of Coca-Cola for more than thirty years, once said that his happiest childhood memory was attending his first baseball game with his dad. And what did he drink during the game? Why, an ice-cold cola, of course. Given that he associated a particular product with one of his fondest memories, Woodruff guessed that others would probably do the same. Armed with this realisation, he tried to increase the chances of that product being Coca-Cola by pursuing a strategy of ubiquity. The idea was to make Coke available wherever there was a reasonable chance of someone experiencing a special, memorable moment in their life – ball parks, seaside resorts, movie theatres, you name it. The strategy worked because the human brain has a tendency to link numerous different elements in a single memory – sights, sounds, emotions . . . and tastes. So Coke became directly associated with millions of happy memories.

Advertisers are constantly attempting to influence our food choices, and all of their strategies rely on neuroscience and psychology. Commercials simply wouldn't work if they didn't affect our brain and the way we think.

Little children are naive. They believe in Santa Claus because their parents say he exists, and they think they should eat Frosties because Tony the Tiger says they're 'gr-r-reat'. I grew up at a time when the public broadcasting company still enjoyed a monopoly on Norwegian television, so I managed to hold on to my naivety for a little longer than my contemporaries in America and the United Kingdom. I still feel embarrassed when I think about how much I spent the first time I tuned in to a shopping channel. Luckily, I soon learned to resist the slick ads and clever sales tricks. However, times are changing. Today's marketing campaigns don't need to adopt the same scattergun approach. If you're a member of a supermarket's loyalty scheme, you're probably already aware that you receive advertising that's targeted specifically at you. For instance, you may receive special offers on cartons of beer and snack foods in the run-up to your local team's appearance in the Cup Final, or details of the store's range of cakes, party bags and decorations just before your son's birthday. It's very difficult to resist this type of targeted advertising, which is why marketing departments spend so much time and money gathering as much information as possible about their customers. How much and what we put on our tables at home becomes a perpetual struggle between the advertising campaigns and our own reasoning, which usually tells us that we don't actually need any of the stuff they're offering. We can gain the upper hand in this battle by reflecting on what influences us and how.

The food industry knows that taste is only a fraction of the overall eating experience. For instance, it has invested millions in making the perfect French macaroon – crispy on the outside but chewy in the middle. And many fizzy drinks would be unrecognisable if the manufacturers dared to meddle with the amount of carbon dioxide they contain. The *feel* of food in our mouths is more important than many of us realise. Similarly, smell alone can be enough to induce a craving. Just the aroma of freshly baked bread is sufficient to make people's mouths water. That's why you'll often smell it as you wander around your local supermarket – in-house bakeries boost sales.

Mouth-watering is no insignificant matter. Saliva production is important because any liquid in the mouth helps the food to reach every single taste bud, resulting in stronger signals to the brain. Food manufacturers are well aware of this, so it's hardly surprising that so many of their products feature sauces and dressings. Similarly, it's not just the combination of fat, sugar and starch in chocolate that we find so appealing but the fact that it melts on the tongue, which promotes the release of saliva, powerful signals to the brain and ultimately the sense of euphoria that dopamine creates.

Food make-up

It's rather strange that we think chocolate *looks* tasty. If it had just been discovered, it's extremely doubtful that the modern food industry would trust us to accept that little brown clumps could be delicious. The manufacturers probably wouldn't be able to resist adding artificial dyes in the hope of stopping consumers' brains from making all sorts of unfortunate associations.

Sweet manufacturers usually take the biscuit when it comes to the use of artificial dyes. They add vivid colours which capture our interest and dangle the tantalising promise of strong flavours. But even people who steer clear of sweets and try to eat healthily can't escape this culinary cosmetics industry entirely. For instance, some bread manufacturers add malt to the dough to make their loaves look more wholegrain and therefore more healthy. Meanwhile, the fish-farming industry has developed a colour chart to boost the sales of its salmon. Wild salmon flesh can be anything from a vivid pink to red because the fish eat shrimp and other crustaceans, whereas fillets of farmed salmon are naturally white. There is supposedly negligible difference in taste, but consumers have learned to associate colour with flavour, so the farmers add the synthetic pigment astaxanthin to their fish feed. As some countries prefer pink and others red, the farmers consult the colour chart and adjust the amounts accordingly, depending on the intended market.

As we've seen, when we eat food we desire, whether it's healthy or unhealthy, we feel a surge of happiness due to the build-up of dopamine in the nucleus accumbens. But before that happens, many other parts of the brain have an influence on *what* we desire. For instance, the amygdala and hippocampus work together so you remember the pleasure you felt the last time you treated yourself to a juicy hamburger or a bag of potato chips, and the insula contributes by enhancing the reward effect. Meanwhile, the frontal lobe puts everything into context and explains that, since you've been so busy and are feeling so tired, you both need and deserve the energy boost that a burger will provide ... or it tells you that you've been eating far too much junk food lately, so you should opt for the salad instead.

The problem with artificial sweeteners

Dopamine isn't the only hormone that floods through your brain when you eat sugar. The appetite-suppressing hormone leptin tells you that you've had enough once you've consumed a certain number of calories. But what happens when you eat something that tastes sweet yet contains hardly any calories? Artificial sweeteners activate the brain's reward system in exactly the same way as sugar, so you feel the dopamine hit; but there's no build-up of calories, so no leptin is ever released. As a result, your brain continues to crave real sugar. This means that artificially sweetened soft drinks aren't necessarily the best option when your sweet tooth rears its head, no matter how many cans you drink, because you'll have to eat some sugar eventually if you want to feel sated.

Chocoholic in the womb?

If your mother ate a lot of garlic while she was pregnant, you've probably liked the taste from an early age. Human foetuses are able to taste and smell substances in the surrounding amniotic fluid from a very early stage of development, and as we become accustomed to certain flavours, we start to like them. For instance, mothers who drink a lot of carrot juice during pregnancy and while they are breast-feeding tend to have children who love carrots. Thus, while people often say they've liked something since the day they were born, in reality they probably developed a taste for it even earlier than that.

After a pregnant woman eats something sweet, the foetus swallows much more amniotic fluid than it does when the mother eats something bitter. And infants who have never eaten anything other than their mother's milk usually adore sugar or sugar-water from their very first taste of it. For instance, those who wake up from anaesthesia and are totally inconsolable will often calm down immediately if their dummy is dipped in a little sugar-water. This does not mean that all crying infants should be pacified with sugar-water, but it illustrates an important point.

It's a very different story with salt. Infants don't like the taste, and they shouldn't be fed salty foods. However, their brains can be trained to tolerate – and even crave – ever larger quantities of it. Our intake of salt has skyrocketed as our consumption of prepared food has increased exponentially over the last few decades, and it has been identified as one of the chief culprits in the prevalence of high blood pressure, heart attacks and strokes in the Western world. People who eat prepared foods for the first time usually find them incredibly salty, but their brains soon get used to the taste. Eventually, they come to expect it, and find meals with ordinary levels of salt bland. Hence, children who are served 'grown-up food' – that is, with a high salt content – will gradually demand ever more salt, while those who are routinely served food with a low salt content tend to avoid excessive saltiness. Fortunately, though, if you force yourself to avoid salt for a while – either as a child or as an adult – you will soon stop missing it.

In contrast to salt, we are preprogrammed to like fat as well as sugar from the very beginning. So, to some extent, we are all chocoholics even before we're born. However, the *amount* of fat we crave is open to influence. For instance, mothers who eat a lot of fatty foods during pregnancy and while breastfeeding tend to have children who need to consume a great deal of fat before their reward centres start releasing dopamine. This means that a mother's antenatal and post-natal nutrition is one of the most important non-genetic factors in a child's brain development.

Brain food

Pregnant women can sing as many lullabies or play as much Mozart as they want for the babies in their tummies, but they should really focus on increasing their consumption of fish, and especially oily fish, because our brains need the type of fat that these creatures have in abundance. The brain is the fattiest organ in the human body, but it doesn't use its fat reserves for energy. Rather, they are utilised in the production of neurons and other cells, especially those that insulate the axons with layers of fat-laden membranes so the signals will travel quickly and efficiently.

There are two main categories of fatty acid: non-essential, which our body can manufacture; and essential, which we can only get from the foods we eat. Several essential fatty acids – including omega-3 – are especially important elements in the brain's physical structure. Omega-3 is found in countless foodstuffs, but we have learned that some of these sources are better than others. The long-chain omega-3 that the brain really needs is most prevalent in oily fish like salmon, trout, mackerel and herring, as well as fish products like cod liver oil. By contrast, while some plant products – such as flaxseed – have high concentrations of omega-3, it is the short-chain variety, and the human body is not very good at converting this into the required long-chain form. So the answer is clear: eat more fish.

Brain development is linked to much more than brain size, but it's difficult to do safe, detailed studies of babies' brains, so head circumference is often used as a proxy measure. A Swedish research team found that infants whose mothers had high levels of omega-3 and omega-6 fatty acids in their milk tended to have larger heads than those who were deprived of these nutrients. Another study found that the children of mothers who supplemented their diet with cod-liver oil (omega-3) during pregnancy and while breastfeeding had larger heads than those whose mothers took corn oil (omega-6) instead. Moreover, when these two groups were tested at the age of four, on average the omega-3 children were found to be more intelligent than the omega-6 children.

But it's not just children who benefit from omega-3. We all need this essential fatty acid to keep our brains in shape. Remember, the brain continues to develop throughout life: new neurons and new neural connections are continually formed, while others die and disappear. Several studies have shown that regular, high consumption of oily fish reduces the risk of developing memory problems in later life, whereas a low level of omega-3 in the blood seems to be a risk factor for Alzheimer's and other forms of dementia.

Diets

It's only when we take the plunge and go on a diet that most of us put some effort into learning a little more about nutrition and exploring what the various products on offer in the supermarket actually contain, and especially how many calories they contain. In terms of losing weight, pretty much any diet will do. As long as you stick to it and consume fewer calories than your body uses, you're guaranteed to shed the pounds because you'll start to convert your reserves of fat into energy. However, you must always meet your minimum energy requirements, because your brain will start to consume itself if your reserves fall too low. This is rarely an issue for people on regular diets, but it is a factor among those who develop serious eating disorders, such as anorexia nervosa.

The type of food from which we obtain our energy is an important consideration for all of us, though. In several popular diets – including Atkins – most of the energy comes in the form of fat. In short, these diets steer people away from carbohydrates but allow them to wolf down as much fat as they want. The idea is to stimulate a process called lipolysis, in which the body meets its energy needs by burning fat, rather than carbohydrates. In the secondary phase of lipolysis – ketosis – the body produces fatty compounds called ketones, which become its principal source of energy. In a normal, balanced diet, ketosis is the body's natural reaction to a crisis situation, so a number of people have claimed that we should not induce it artificially. On the other hand, many studies have found that the adult brain continues to function just fine on ketones and suffers no lasting damage. The only problem is that you don't enter ketosis as soon as you start the diet, which can leave the brain temporarily starved of energy; and if you don't have much energy, you tend to do poorly in intelligence tests. However, this normalises after you've been on a low-carb diet for a while.

Crucially, though, all of this only applies to adults. Developing brains are another matter entirely, especially during pregnancy, when the foetus's neurons are forming. In experiments with rats, if the mother is fed a fat-laden diet, the foetus produces more neurons in the area of the brain that regulates hunger. Consequently, after birth, the infant mouse has a larger appetite, prefers fat, has a high level of fatty acid in its bloodstream and a tendency to become overweight.

When it comes to food, we must remember that the primitive part of the brain rewards the consumption of all sources of energy, without any consideration of the long-term consequences. Fortunately, though, the cerebral cortex can override the primitive reward centre, so we aren't forced to gorge on fat and sugar until we suffer an early heart attack or stroke. Eat everything in moderation, and be extra careful if you fall pregnant. After all, in the course of those nine months, you will have responsibility for the development of *someone else*'s brain.

Chapter 9

Addiction

After the long, dark Norwegian winter – when a hot cup of coffee always feels like an essential start to the day – it always seems sensible to reset my system. So, every summer, I abstain from drinking coffee for a whole month. Thereafter, I limit myself to an occasional cup, which has an instant impact and helps me feel more awake and focused. However, as the autumn draws to a close, I feel the need for more and more coffee, until I'm back to starting every day with a cup. Before long, I wake up each morning with what seems like a deficit – I need the coffee just to make it back to zero. And if I've had trouble sleeping, one cup just won't cut it; I need two. This is an example of dependency and addiction.

Every substance you consume that has a direct impact on your brain can be defined as a drug. According to this definition, coffee is the most widely used drug in the Western world. It stimulates your nervous system, so it is known as a central nervous system stimulant. Others include cocaine, amphetamines and nicotine. Substances that depress your brain activity are known as central nervous system depressants. The most common of these is alcohol, but heroin and cannabis are in the same category.

Dependency

What causes a person who was once a little darling to start stealing from her parents to get her next fix? There are many natural motivation and reward systems in the brain, and they reward us whenever we achieve specific goals. In addition, humans have learned that smoking, eating, drinking or injecting certain substances can trigger the same reward systems without any need to reach those goals. It's cheating, but most of us do it to some extent.

Addictive drugs have an impact on the brain's chemistry in three ways: they can resemble one of the human body's natural neurotransmitters and block its path to the brain's receptors; they can stimulate the release of one or several of those neurotransmitters; or they can prevent the reabsorption of a neurotransmitter into the neurons that released it. When the brain is affected by one of these substances, it tries to regain balance by activating its defence mechanisms. For instance, after a drug has stimulated the release of dopamine for a certain period of time, the brain will start to shut down some of the receptors that are available to that hormone. This reduces the overall efficiency of the brain's reward system, so you experience less of a dopamine rush from sex, food or exercise. You also need to smoke, snort or drink more in order to enjoy the kind of high you have come to expect. In other words, both natural and artificial highs are harder to achieve. This is called 'tolerance', and it is due to physical changes in the brain.

Therefore, what people call psychological addiction has a physical dimension, too. An example of the psychological aspect is when a smoker feels her stress melting away the instant she places a cigarette in the fingers of her right hand – before she's even lit it. Many smoking cessation methods try to break this kind of mental association, for instance by encouraging smokers to hold the cigarette in their 'wrong' hand, which generates much less expectation of relaxation and pleasure. Some of these techniques have proved highly successful, but the physical addiction has to be addressed, too.

By definition, the neural networks that transmit the signals relating to habitual behaviour are activated time after time, so they tend to be extremely strong and stable. However, you can override them. For instance, the neural networks relating to smoking will eventually deteriorate if a smoker repeatedly resists the temptation to reach for a cigarette every time she feels stressed and instead learns to handle the tension in some other way.

But the best way to avoid addiction is never to start!

Coffee

The first time I decided to give up coffee for a month, immediately after my final exams, I tried to do it cold turkey and spent the next two days with a splitting headache. I should have known better.

Because caffeine resembles the neurotransmitter that helps you feel tired, and blocks its path to the receptors that it usually acts upon, you feel more awake and alert for the first few hours after you drink a cup of coffee. In addition, some of the brain's other neurotransmitters, including dopamine, work more effectively when the one that makes you feel sleepy is kept away from your receptors. This prompts your adrenal glands to release adrenaline, which heightens your alertness and sharpens your focus. All of this can come in handy if a sick child has kept you up all night and you have to give a presentation at work the next day. However, if you try to replicate the trick every day, you should know that your brain will soon start to make allowances for your behaviour. Once the sleep-inducing neurotransmitter has been blocked a certain number of times, your brain will set about making more receptors for it. As a result, if you keep drinking the same amount of coffee, you will end up feeling just as tired as you did before. So begins a vicious circle: to experience that familiar coffee pick-me-up, you simply have to drink more . . . and more . . . and more. On the other hand, if you stop drinking coffee altogether, you will put yourself in a deficit situation. With no more caffeine blocking any of your receptors, the sleep-inducing neurotransmitter is suddenly able to act on both the new and the old receptors, which will leave you feeling not only tired but absolutely shattered. This is an indication that you're addicted to caffeine. If your habit has spiralled to the point where you need numerous cups of coffee each day just to function normally, then you're sure to have a correspondingly large number of receptors for the sleep-inducing neurotransmitter, so you'd be wise to reduce your intake gradually instead of going cold turkey.

Every time I give up coffee, I comfort myself with the knowledge that my receptors will start to normalise in a week or so. Nevertheless, even once I'm well beyond that point, whenever I smell the aroma of freshly brewed coffee, there's nothing I want more. That's a manifestation of the learned aspect of addiction and it's harder to shake than the physical craving. In this case, my neural pathways are telling me that I 'need' coffee because my alarm clock went off so early.

Even when I'm not in a caffeine-free month, I make it a rule never to drink coffee after lunch. It's not that coffee stops me from sleeping (I'm one of those people who can fall asleep anywhere at any time), but I know that caffeine continues to affect the body for much longer than you might think. For instance, if you drink a cup of coffee at lunchtime, about 25 per cent of the caffeine will still be in your system at 10 p.m. While that's not enough to prevent me from falling asleep, it will reduce the *quality* of that sleep. And that will mean I'll need even more coffee the next day.

Cocaine and amphetamines

Cocaine is a natural substance, derived from the coca plant, while amphetamines are manufactured synthetically. Like caffeine, they are both central nervous system stimulants, but while caffeine resembles and therefore blocks the neurotransmitter that makes us tired, cocaine and amphetamines alter the quantities of some of the body's other hormones. Cocaine increases the levels of the fight-or-flight neurotransmitter noradrenaline and the reward hormone dopamine by preventing their reabsorption into the neurons that have released them. Similarly, amphetamines (and methamphetamines) increase the level of dopamine. However, it's not just the increased level of dopamine that's important, but where it occurs. We have already seen that the nucleus accumbens in the basal nuclei releases dopamine naturally as a reward for achieving some sort of goal or target, such as providing the brain with much-needed energy (see Fig. 20). Cocaine's and amphetamines' effectiveness rests on the fact that they trick the nucleus accumbens into issuing a dopamine reward for something as insignificant as taking a drink of water when you're thirsty. Consequently, if you take one of these drugs, you are overwhelmed by the reward sensation, even though these substances are bad for you.

Put simply, noradrenaline makes you more alert and dopamine makes you happier – at least initially. Gradually, though, your brain will become desensitised to both of these hormones, so the things that once gave you pleasure or triggered your excitement will leave you cold. Ultimately, if you become an addict, only cocaine will make you happy.

Nicotine

After each drag on a cigarette, it takes only ten seconds for the nicotine to reach your brain. Once it's there, it replicates the action of the neurotransmitter acetylcholine on the neurons of the limbic system, which promotes the release of the reward chemical dopamine. So it's actually the hit of dopamine that smokers crave when they think they're longing for a cigarette. Further afield, nicotine also prompts the adrenal glands to release the stress hormone adrenaline. Hence, it is considered a stimulant. Nevertheless, if smokers don't have access to cigarettes, they start to feel more, rather than less, stressed and agitated because their brains are deprived of nicotine's sedative effects. That's why, paradoxically, smokers often say they need a cigarette to get going in the morning but also need one to relax before going to bed at night.

Smoking tobacco has undoubtedly shortened the lives of millions of people who have died from cancer, heart attacks or strokes. Yet, in light of its powerful impact on the brain, a number of studies have explored the use of nicotine in the treatment of Parkinson's disease and Alzheimer's. Some of the results have been encouraging. This is just one of many examples of a toxin having a positive effect if it is administered correctly.

Alcohol

Alcohol has some influence in most regions of the brain because it binds to the receptors for a variety of neurotransmitters, including those for serotonin, which helps explain its numbing effect. Hence, under normal circumstances, communication between the neurons slows to a crawl under the influence of alcohol. But if the brain grows accustomed to alcohol due to long-term use, it tries to increase the signal speed between the neurons by releasing activating neurotransmitters. So, if an alcoholic suddenly stops drinking, their brain is left with far too many of these activating neurotransmitters, which continue to stimulate it until it's totally out of control. This is why it's so dangerous to quit cold turkey if you drink large quantities of alcohol. You could have vivid hallucinations or even suffer seizures.

Alcohol would probably be banned immediately if it were discovered today. Foetuses with alcoholic mothers suffer more harm than those with mothers who are heroin addicts. That's because alcohol can cause serious, irreversible brain damage in unborn children. Moreover, in spite of dozens of studies, nobody has been able to establish a definitively safe low limit for alcohol consumption during pregnancy.

Of course, alcohol can also cause brain damage in adults. And this problem is probably more widespread than you realise. In the Western world, a good 10 per cent of the population will meet the criteria for alcoholism at some point in their lives (usually under the age of twenty-five). When a heavy drinker starts to experience jerky body movements, eyes that droop to the side, memory problems and confusion, they might be in the early stages of Wernicke–Korsakoff syndrome, a condition in which the brain shrinks due to vitamin B_1 deficiency. The regions of the brain that are most affected are two parts of the limbic system that resemble a pair of breasts – the 'mammillary bodies' – the thalamus and the white matter. The cerebral cortex shrinks too, but that's probably due to the toxic effect of the alcohol itself, rather than the shortage of B_1. Alcoholics lack B_1 because alcohol prevents the vitamin's absorption in the intestines, as well as its storage and conversion into the usable, active form in the liver. Deficiency is extremely serious because B_1 facilitates the brain's use of blood sugar for energy, and it also plays a role in the production of some neurotransmitters and myelin, the neurons' insulation material.

People who are under the influence of alcohol stagger because of its effect on the cerebellum, and they lose their inhibitions because of its effect on the frontal lobe. For instance, after a few drinks, you might not feel any of your usual fear of rejection when seeing someone attractive in a bar. You will simply wander – or stagger – over and ask them back to your place. Moreover, this lack of fear is usually accompanied by a heightened interest in sex, again because the frontal lobe is unable to perform its usual constraining function. Unfortunately, though, alcohol has another couple of tricks to play. It inhibits the centres in the hypothalamus and the pituitary gland that control sexual function, so while your brain might want sex, your body might not be so keen (see Fig. 18). And if you pass out before the act is finished, you can blame that on the booze too, because you feel sleepy when the brainstem is flooded with alcohol.

The long queues for the bathrooms in your local night-club are also a consequence of alcohol consumption. The pituitary gland usually releases a hormone that causes your body to retain liquid so you don't get dehydrated. However, alcohol inhibits the release of this hormone, so you feel the urge to pee more. As a result, you dehydrate, which is one of the chief causes of the splitting headache you suffer the next day. Your brain actually shrinks because of the water loss, so it tugs painfully on the surrounding membranes.

There's more to a hangover than just a headache, though. When you pee too often, you excrete many of the salts that play important roles in the transmission of nerve signals and muscle control. You also become nauseated and exhausted, which is exacerbated by poor-quality sleep. Some people drink to help them fall asleep, which seems logical enough, as alcohol has a sedative effect and inhibits the brain's activating neurotransmitters. Indeed, everyone falls asleep eventually if they drink enough alcohol. However, when you stop drinking, your brain compensates by producing more of the activating neurotransmitters than it needs. This prevents you from falling into a deep, rejuvenating sleep. Moreover, the overproduction of activating neurotransmitters can lead to uncontrollable trembling, a sense of uneasiness and even increased blood pressure the following day. Finally, alcohol is absorbed directly through the stomach wall and contributes to the formation of hydrochloric acid. If you produce too much of this acid, the nerves around your stomach will send a warning signal to your brain that your body is under threat, and your brain will respond by initiating your vomiting reflex.

How hungover you feel also depends on what you were drinking the night before. Hangovers are milder if you stick to clear liquids, such as white wine or vodka, because coloured drinks, such as red wine and tequila, contain additional toxins, like tannic acid. Of course, if you want to avoid a hangover altogether, it's best to avoid alcohol completely. If that's not an option, you should at least drink a glass of water for every alcoholic drink, as that will counteract the effects of dehydration.

Endorphins, morphine and heroin

Endorphins are the brain's own narcotics. These neurotransmitters are released when you experience stress or feel pain, and they are particularly active in the limbic system. They are released during exercise – and childbirth – to give you a natural high. Yet it is impossible to get addicted to them. Like other hormones, they are released into the spaces between neurons and then received by the receptors on the other side of the gap. But the instant an endorphin binds to one of these receptors, it is broken down and recycled.

Morphine and its chief derivative, heroin, are chemical compounds that closely resemble endorphins and therefore fit neatly into the brain's endorphin receptors (see Fig. 10). However, unlike endorphins, they aren't immediately broken down, so they continue to activate the receptor over and over again. Other drugs have a similar effect, but heroin and morphine resemble a number of different neurotransmitters, so they affect a wide variety of receptors. The brain tries to normalise this overstimulation by gradually reducing the number of functional receptors. As a result, a heroin addict needs to use more and more to feel the same effect. Similarly, a patient who receives regular morphine injections will become habituated and eventually experience no pain relief from the drug without an increase in the dose. Moreover, if the morphine or heroin is suddenly withdrawn, the brain's endorphins will be unable to compensate because of the shortage of functional receptors. This results in restlessness, muscular pain, sleeplessness and nausea. Fortunately, all of these withdrawal symptoms cease when the number of receptors returns to normal a few weeks later, although the psychological addiction remains. Even a fully functioning endorphin system cannot provide the same high as heroin.

While heroin is less harmful to foetuses than alcohol, neither it nor morphine is harmless. Heroin is particularly detrimental to the brain's white matter, which affects addicts' decision-making ability, stress regulation and overall behaviour. However, its most obvious and immediate effect is due to the fact that it sedates the brainstem's breathing centre at high doses. This results in serious, life-threatening injuries because the brain and the body's other organs are starved of oxygen.

The possession of heroin (let alone its production or distribution) is illegal in all but a handful of countries around the world, whereas doctors routinely prescribe morphine for pain relief. Most countries require medications that can impair an individual's ability to function normally – such as morphine – to print a large warning on the packaging. Many different categories of legally prescribed drugs bear this warning, but it's important to remember that most of these medicines are also addictive, and that they can be abused in much the same way as heroin. Indeed, there are more fatal morphine overdoses each year than fatal heroin and cocaine overdoses combined.

Hash

Hash and marijuana – which are both derived from the cannabis plant – resemble neurotransmitters called endocannabinoids. As we have seen throughout this book, most neurotransmitters are released from neuron number one and travel across the synaptic gap to neuron number two. Endocannabinoids are different because they travel from the neuron that is usually the recipient to the one that is usually the sender. Some neurotransmitters activate and others inhibit the brain's neurons, and a signal won't be sent unless the former outnumber the latter. Endocannabinoids help this process by inhibiting the inhibitors. They affect your mood via your amygdala, your memory via your hippocampus and the general functioning of your cerebral cortex. Cannabis overstimulates the receptors for natural endocannabinoids in all of these regions of the brain, which leaves them unable to do their job of regulating the signals that the neurons send between themselves. Given that cannabis's zone of influence is so wide, it's hardly surprising that it has multiple effects – from altering the user's perception of time to leaving them feeling relaxed, giddy or euphoric, and from inducing panic and paranoia to impairing concentration, learning and memory. In rare cases, it may even result in acute psychosis.

We know that foetuses who are exposed to hash tend to become overly impulsive as they grow up, and they can also develop learning difficulties and memory problems. Thus, hash clearly damages the young, developing brain. Yet many users insist that smoking weed causes no lasting harm to the adult brain. At least one group of researchers disagree, because they found that the use of hash greatly increases the risk of developing schizophrenia.

Schizophrenics struggle to distinguish between their own imagination and the real world. They are delusional and may see hallucinations and hear voices. 'Schizophrenia' is used as a loose, umbrella term for a number of different brain disorders that manifest in a variety of symptoms. However, all of these disorders are chronic: that is, once you develop one of them, you will have to live with it for the rest of your life. The aforementioned research project found that 10 per cent of the hash smokers in the study group developed some form of schizophrenia, as opposed to 3 per cent of the general population. Now, it should be pointed out that schizophrenia can be triggered by a wide variety of factors, so it's impossible to say for certain that the drug was entirely to blame. But would you really want to run the risk of tripling your likelihood of developing it?

The marijuana lobby also often argues that it is not addictive. However, studies have shown that up to 10 per cent of regular smokers become addicted (compared to 20 per cent of heroin users). This should come as no surprise because all addictive drugs have a much stronger influence on the brain's reward system than natural stimulants like food, sex or exercise. The brain learns that it can gain trigger euphoria – or gain some respite from difficult situations and troubling emotions – through intoxication. Just as Pavlov's dogs started yearning for food when they heard a bell, smokers crave a cigarette after dinner because they know it will heighten their pleasure and contentment. And the same is true of alcoholics who yearn for their next drink, and heroin addicts who long for their next fix. All addicts build new synapses between the brain's neurons as well as whole new neural networks, and these strengthen the urge to smoke, drink or shoot up. And the networks don't disappear overnight when you quit. Indeed, some of them remain for the rest of your life as unwelcome reminders of what you've given up.

Chapter 10

Reality versus Perception

Look around. Do you think you perceive the world precisely as it actually is? In the movie *The Matrix*, Neo is given a choice between taking a blue pill and remaining in a dream world, or taking a red pill and following a white rabbit down a hole to the real world. He chooses the red pill, but soon learns that, in many ways, the dream world is far preferable to reality. There's certainly much less chance of imminent death.

Since antiquity, philosophers have pondered whether what we feel, smell, taste, hear and see is real. If it is, then reality is just a series of electrical and chemical signals that our brains interpret in a particular way. But what alternative do we have, because we only have access to the physical world via our senses?

Thinking about this leaves you feeling a little like Neo in *The Matrix*. You realise that the brain only ever provides an adapted image of the world, and that what the human eye sees isn't some sort of universal, absolute reality. Rather, the brain uses the information from our senses to draft its own personal interpretation of the world. We call this 'perception'.

Your amazing sense of smell

As we saw in Chapter 3, our sense of smell plays a crucial role whenever we choose to remember a particular incident. But it's also often the first sense to react to a dangerous situation. For instance, it warns us of the threat of fire long before we see the flames; and it makes us gag or even vomit before we have a chance to eat rotten food. So what we call 'smell' comprises much more than a group of molecules making their way up our noses. In addition to eliciting a powerful memory, a familiar scent can prompt us to set higher goals for ourselves or find more effective solutions for a particular task. You can test this by standing outside a bakery. Studies have shown that people who are exposed to the scent of freshly baked bread (or freshly brewed coffee) are more inclined to help strangers. This seems to be because we associate lovely smells with good memories, and good memories put us in a positive frame of mind.

Of course, some people will make very different associations. For example, if, as a child, you gorged on freshly baked bread until you were violently ill, every time you smell it now, your mood will probably worsen, not improve. On the other hand, researchers have discovered that the children of mothers who smoked, drank alcohol or ate a lot of garlic during pregnancy find those smells appealing. In other words, when our brain receives and interprets olfactory information relating to a particular smell, it does not provide an objective impression of reality; rather, it makes a subjective assessment on the basis of personal experience.

Fooling your taste buds

'Wow, that was so good!' exclaimed the biology student who ate a whole lemon after first sucking on a tablet containing an extract from the so-called 'miracle berry'. The lemon tasted so delicious because a certain protein from the berry had bonded to the cells on his tongue. Then, when he reduced the acidity in his mouth by eating the lemon, the protein was activated and sent signals to his brain that he was eating something sweet – even though he wasn't.

All of our senses can be tricked, which can distort our image of the world around us. The food industry knows this and fools our sense of taste by inhibiting bitter aftertastes and promoting sweet flavours, and it will probably find a host of new ways to do so in the future.

The taste of crunch

We've already seen that taste is closely linked to smell. If you've ever had a blocked nose, you've surely eaten a meal that seemed to have no taste whatsoever. But taste is connected to more than just smell. The way food feels in your mouth and even the sounds you hear when you bite into it both have an influence on how it tastes.

Whenever I went to the movies as a child, my parents would insist on decanting my crisps from their packet and into a soft plastic bag, so I wouldn't disturb the other cinema-goers. The experience just wasn't the same. The rustling of the packet generates an expectation of what's to come: namely salty, crispy, fried slivers of potato. By the same token, soggy day-old crisps are no fun either, even though the taste and the smell are exactly the same. On the other hand, research has shown that I would have perceived the crisps to be even more delicious than usual if the sound of their crunch had been amplified.

The taste of red

In addition to smell and sound, taste depends on sight. That's why sweets are often dyed vivid colours: the colour stimulates expectation. In an experiment, a group of children was given identical-tasting red jelly beans and yellow jelly beans. The children generally thought that the yellow bean was sour and the red bean was sweet, based solely on their respective colours.

So, our sense of taste is not objective, either. Rather, the brain gathers information from our senses of smell, touch, hearing and vision then combines that with the signals it receives from our tongue before assigning a particular taste to anything.

What you don't feel

Your skin is full of receptors that collect information about where your right hand is currently located, or how it feels to wear a ring on your finger. If you're not used to wearing a ring, at first you'll be constantly aware that something unfamiliar is wrapped around your finger. That's why newlyweds often fidget with their rings. Eventually though, as you get used to it, your brain filters out all of the somatosensory (sense of touch) information relating to the ring.

It's like that with everything. Your brain receives constant signals from your skin about the clothes you're wearing, the chair you're sitting on, the hairs that are touching your forehead and the book you're holding in your hands. But if you had to respond to this information all the time, you wouldn't be able to concentrate on reading these words, so your brain filters out all the superfluous, unimportant details. In other words, your perception of reality is censored by your own brain.

Selective hearing

What we call 'sound' is actually just a series of variations in the air pressure next to your eardrums. Moreover, we only hear those pressure variations within a certain – quite limited – range. So what seems like a quiet room to us might sound like a cacophony to a mouse. Our ears are specifically attuned to hear other people and imminent danger. That's enough. Just think how unbearable it would be if they could detect every single sound.

We don't really hear anything until the brain interprets the ears' signals relating to those changes in air pressure. So you need your brain in order to listen to music or chat with friends. For instance, it tells you that the sounds that are coming from your friends' mouths are language. If we hear a word in a language we understand, it's more than just a sound to us: it has meaning. And we interpret that meaning regardless of the quality of the sound or the amplitude and frequency of the sound waves. That is, we understand a word regardless of whether it was spoken by a deep or a high voice, and regardless of whether it was screamed or whispered.

If the brain didn't filter out unimportant sounds, we would all want to live like hermits. Certainly, no one could bear to live in a big city or next to a busy road. And no one would ever visit a shopping mall. Yet, because your brain's filter is so efficient, you can conduct an intimate conversation even while doing your Christmas shopping. At least until you hear a familiar voice calling your name. And that particular sound doesn't even need to be loud, because your brain will recognise its importance immediately and let it straight through the filter and into your consciousness.

A world without depth and contrast

How many muffins could you make in this muffin pan?

Do you see the top one in the middle as sticking up instead of down? If you invert this book and look at the picture again, you'll see it the other way around – the one in the middle at the bottom now seems to be concave rather than convex.

Your vision has evolved to help you recognise objects, and even though the images on your retinas are flat, your brain helps you see the world in three dimensions. However, its visual cortex expects just one light source to illuminate everything you see – shining down from above, like the sun or the moon, which would cause three-dimensional objects to cast familiar shadows. Hence, although the muffin pan picture is two-dimensional, your brain sees the familiar shadows and automatically interprets each muffin cup as either a three-dimensional convex shape (curving up, like a bump) or a concave shape (curving down, like a hollow). Again, your brain interprets the information it receives to generate what it believes to be the most accurate version of reality, but sometimes it can be fooled.

Which of these squares is darker?

You probably think the square on the right is darker, because your eyesight automatically increases the contrast with the background. That is, your brain steps in to help you see more. Actually, the two squares are exactly the same shade of grey, which will become apparent if you cover the background so that you see only the squares themselves.

Infrared and ultraviolet light

The picture of the world that your brain paints for you is slightly different from the one that other people see, and radically different from the one that other animals see. If nocturnal animals' senses were as bad as ours, they would starve to death in a week. Several snake species, for example, have infrared sensors that allow them to hunt at night. Meanwhile, bees can see ultraviolet light, which allows them to perceive nectar guides on the petals of flowers.

Facial recognition

Do you see what's wrong with this picture?

When we look at faces, we usually recognise what they're expressing almost immediately. We do this by focusing primarily on the eyes and mouth. That's how we work out if someone is feeling angry or happy. In this (mostly) inverted picture of Professor May-Britt Moser, her eyes and mouth are the right way up, so it was probably a few seconds before you noticed anything untoward.

Now invert the book and look at the picture again. The most important parts of her face – her eyes and mouth – are now upside down, so you immediately see that something is wrong. You realise that your brain took a shortcut and overlooked a few crucial details. It used cognitive reasoning and assumptions about the visual information it received to save time and energy and create a tailored picture of the world.

These simple visual experiments prove that we don't always see the world how it actually is. Rather, our brains ensure that we see a refined – but more useful – version of it.

Why perception is better than reality

Indeed, the brain interprets the signals from *all* our senses and then delivers the results to our consciousness in the most useful way. These results aren't necessarily exact representations of how the world actually looks, sounds, smells, tastes or feels at that precise moment, but rather how it *should* look, sound, smell, taste or feel. Through this process, our brain helps us understand what we're seeing, hearing, smelling, tasting and feeling. It helps us *perceive* the world around us.

It's perception that allows us to convert decibels and hertz into music. It's the reason why flowers smell good and sewage farms smell bad. It's the reason why chocolate is delicious and mouldy food turns our stomach. Perception is *better* than reality. But, in another sense, our perception *is* our reality. While you can never be quite sure that your brain's interpretation of the information it receives from your senses is totally accurate, you can rest assured that it creates the best reality for you.

Without the brain's gift of perception, we wouldn't understand art. In fact, there would be no culture of any kind. We would live in a world without music because sound would be nothing more than continual variations in air pressure.

Chapter 11

The Way Forward

In this book, we have seen how important the brain is to our species' success and that, fundamentally, we *are* our brain. The brain makes it possible for us to love, but it can also make us afraid or jealous. All of our thoughts and emotions are *physical* processes in the brain – the result of signals sent by neurons through their networks.

Human intelligence is also a consequence of how our brain is constructed and how our neurons communicate with one another. This is true regardless of whether you measure intelligence in terms of IQ or use psychologist Howard Garner's definition, in which it is divided into various types, such as linguistic, musical or social intelligence.

Learning is physical too, because the structure of the brain changes as we learn. And it is surprisingly flexible. We can learn to seek solace in drugs, alcohol or unhealthy food, but we can also learn to speak new languages or find our way around in new places. May-Britt and Edvard Moser are continuing their research in this area because they know that we have barely touched the surface so far.

While I hope that this book has provided many answers, some of the questions I posed at the beginning remain unanswered. Where do thoughts begin? What is free will, and do we really possess it? In addition to these philosophical questions, there are others that relate to more pressing issues. What causes Alzheimer's disease? And can we do anything to prevent its progression?

One in three Westerners will be afflicted by a disease or injury to their nervous system at some point in their life. This makes it the most important cause of illness in the Western world. But we can't understand the big picture of brain diseases without first understanding the intricate workings of the brain itself. Scanning the brain of a person who is suffering from depression won't provide all of the answers about what is causing their illness. Indeed, if we were to define depression on the basis of its causes, rather than its symptoms, we would probably discover that it's actually a combination of many different brain disorders. So, if we want to understand an illness like depression, we need to start looking at the small – microscopic – picture, rather than the big picture. My work concentrates on the small picture, but it will eventually bring the big picture into sharper focus. By figuring out how neurons communicate with one another, we are moving closer to understanding the brain and developing the tools we need to tackle conditions like epilepsy, depression and Alzheimer's.

I sometimes feel jealous when my colleagues are able to relate their research projects directly to well-known diseases that affect millions of people: 'I'm working on solving the mystery of cancer,' they might say. By contrast, I have to explain why it's important that N-acetylaspartylglutamate is released from postsynaptic vesicles in excitatory synapses. It's tough, but it's also challenging and exciting. Brain and nervous system diseases cost society as much as cardiovascular disease, cancer and diabetes put together. So funding research into what causes these diseases is money well spent.

Moreover, as our knowledge of the brain continues to grow, we will not only develop better treatments but also learn more about who we are and how the human mind works.

Increasing our understanding of this fantastic organ demands both clinical research into brain diseases and further neuroscientific research into how it performs its routine functions. In the future, if doctors, psychologists and neuroscientists join forces and work together, we should finally start to see the big picture and your superstar brain will be revealed in all its glory.

Acknowledgements

This book would never have come about without the wonderful people at the University of Oslo's Institute for Basic Medical Research, the Neurosurgery Department at the University of Oslo National Hospital, the Neurology Department at Akershus University Hospital, everyone at Kagge Publishing and my ever-supportive family. I thank them all, but I would like to take this opportunity to mention a few of them by name.

I grew up in a home where the natural urge to explore and investigate was encouraged. So I have to thank my parents, Grete and Bjørn, for letting me know that I could achieve whatever I desired from the very beginning. I also wish to thank many people in the scientific world for encouraging my curiosity and instinct to explore. In particular, I would like to mention Emeritus Professor Jon Storm-Mathisen, former chair of the Synaptic Neurochemistry Laboratory at the Institute of Basic Medical Sciences and the Anatomy Department, and neurologist and senior researcher Vidar Gundersen for placing their trust in me when I first walked into the lab at the age of nineteen in the hope of giving brain research a try. As a dissertation adviser, Vidar was also a knowledgeable sparring partner for the duration of my research work. The person with whom I have collaborated most closely during my subsequent research is Associate Professor Cecilie Morland. Although she was far more experienced than me when we met, she has always treated me as an equal partner in all our projects. She has made the grey days lighter, and the light days brighter. I also wish to thank Professor Tormod Fladby at the Neurology Clinic, Akershus University Hospital, for welcoming me into his research team. I look forward to taking that research a step further in the future and making a contribution to solving the mysteries of Parkinson's disease.

With respect to this book, I especially want to thank my little sister Guro, who drew the illustrations for the book. I've always known she is a highly skilled artist, but I never dreamed that she would take on the role of illustrator alongside her principal responsibilities as a project manager in the healthcare sector. So I feel incredibly lucky that she volunteered. She is the one person I know who is even more of a perfectionist than I am myself, so I never felt as if I was being too pernickety when we discussed the illustrations. Moreover, in addition to creating all the images, she read the book chapter by chapter and gave me tremendously valuable feedback. Thank you so, so much, Guro! My mother and my other sister, Birte, also read the manuscript, and Birte edited the captions for all the illustrations and photos. I definitely lucked out with my whole family!

For scientific guidance, I was fortunate to secure the help of Professor Emeritus Leif Gjerstad, from the Neurology Department at Oslo University Hospital's National Hospital. His feedback has been incredibly helpful. Clinical nutritional physiologist Christine Gørbitz and clinical consulting neurologist Are Brean, who is also editor-in-chief of the *Journal of the Norwegian Medical Association,* helped with the chapter 'Eating with Your Brain'. Many thanks to both of them.

I also wish to express my sincere gratitude to Nobel Prize-winner May-Britt Moser for agreeing to write the Foreword. In a field that has traditionally been dominated by men, she is a wonderful role model for the next generation of brain researchers. I am tremendously honoured that she has been so supportive of this project.

I was fortunate to receive both a first-time author's grant and a grant to popularise the natural sciences from the Norwegian Non-Fiction Writers and Translators Association (NFF) as well as a grant from the Fritt Ord Foundation to write this book. Karen Agnes Inglebæk Thue, whom I met at Kagge Publishing's autumn party in 2015, came up with the book's original Norwegian title. I also want to thank Kagge Publishing and my editor Guro Solberg for the invitation to write this book, and all the help they provided to turn it into a product of which we can all be proud.

Of course, I want to thank my husband, Carl Christian, for supporting me in all my endeavours, no matter how many balls I have in the air. And finally, I need to thank our little daughter, Aurora, as she has given me a personal perspective on everything I know, and everything I am still learning, about brain development.

Kaja Nordengen, summer 2016

Selected Sources

I have not included sources for any information so established that it is found in textbooks. New or less well-known research, however, is included in the list of sources below.

Thought (R)evolution

Azevedo, F. A. C., et al., 'Equal numbers of neuronal and non-neuronal cells make the human brain an isometrically scaled-up primate brain', *Journal of Comparative Neurology* 531.5 (2009): 532–541.

Herculano-Houzel, S., et al., 'Cellular scaling rules for rodent brains', *Proceedings of the National Academy of Sciences* 103.32 (2006): 12138–12143.

Herculano-Houzel, S., et al., 'Cellular scaling rules for primate brains', *Proceedings of the National Academy of Sciences* 104.9 (2007): 3562–3567.

Li, H. and Durbin, R., 'Inference of human population history from individual whole-genome sequences', *Nature* 475.7357 (2011): 493–496.

Hunting for the Personality

Ferraris, C. and Carveth, R., 'NASA and the Columbia disaster: decision-making by groupthink?', *Proceedings of the 2003 Association for Business Communication Annual Convention.* (2003).

Haggard, P. 'Human volition: towards a neuroscience of will', *Nature Reviews Neuroscience* 9.12 (2008): 934–946.

Henningsen, D. D., et al., 'Examining the symptoms of groupthink and retrospective sensemaking', *Small Group Research* 37.1 (2006): 36–64.

Janis, I. L., *Groupthink: Psychological studies of policy decisions and fiascos,* 2nd ed., Boston: Houghton Mifflin (1982).

Sperry, R. W., 'Consciousness, personal identity, and the divided brain', Frank Benson, MD & Eric Zaidel, Ph.D.(Eds.) *The Dual Brain* (1985): 11–27.

Strayer, D. L., et al. 'A comparison of the cell phone driver and the drunk driver', *Human factors: The journal of the human factors and ergonomics society* 48.2 (2006): 381–391.

Vestly, A-C., Lillebror and Knerten, *Gyldendal Norsk Forlag AS* (2012): 13.

Memory and Learning

Black, J. E., et al., 'Learning causes synaptogenesis, whereas motor activity causes angiogenesis, in cerebellar cortex of adult rats', *Proceedings of the National Academy of Sciences* 87, no. 14 (1990): 5568–5572.

Bliss, T. and Lømo, T., 'Long-lasting potentiation of synaptic transmission in the dentate area of the anaesthetized rabbit following stimulation of the perforant path', *The Journal of physiology* 232.2 (1973): 331–356.

Corkin, S., 'What's new with the amnesic patient HM?' *Nature Reviews Neuroscience* 3.2 (2002): 153–160.

Cowan, N., 'What are the differences between long-term, short-term, and working memory?', *Progress in brain research* 169 (2008): 323–338.

Depue, B. E., et al., 'Prefrontal regions orchestrate suppression of emotional memories via a two-phase process', *Science* 317.5835 (2007): 215–219.

Elbert, T., et al., 'Increased cortical representation of the fingers of the left hand in string players', *Science* 270, no. 5234

(1995): 305–307.
Fields, D. R., 'White matter in learning, cognition and psychiatric disorders', *Trends in neurosciences* 31.7 (2008): 361–370.
Hassabis, D., et al., 'Patients with hippocampal amnesia cannot imagine new experiences', *Proceedings of the National Academy of Sciences* 104.5 (2007): 1726–1731.
Herz, R. S. and Engen. T., 'Odor memory: review and analysis,' *Psychonomic Bulletin & Review* 3.3 (1996): 300–313.
Molinari, M., et al., 'Cerebellum and procedural learning: evidence from focal cerebellar lesions', *Brain* 120.10 (1997): 1753–1762.
Nabavi, S., et al., 'Engineering a memory with LTD and LTP', *Nature* (2014).
Owen, A. M., et al., 'Planning and spatial working memory following frontal lobe lesions in man', *Neuropsychologia* 28.10 (1990): 1021–1034.
Packard, M. G., and Knowlton, B. J., 'Learning and memory functions of the basal ganglia', *Annual review of neuroscience* 25.1 (2002): 563–593.
Parker, E. S., et al., 'A case of unusual autobiographical remembering', *Neurocase* 12.1 (2006): 35–49.
Proust, M., *Veien til Swann 1, På sporet av den tapte tid,* translation by Anne-Lisa Amadou, Gyldendal Norsk Forlag (1984): 59–63.
Quiroga, Q. R., et al., 'Invariant visual representation by single neurons in the human brain', *Nature* 435, no. 7045 (2005): 1102–1107.
Rolls, E. T., 'The orbitofrontal cortex and reward', *Cerebral Cortex* 10.3 (2000): 284–294.
Scoville, W. B. and Milner, B., 'Loss of recent memory after bilateral hippocampal lesions', *Journal of Neurology, Neurosurgery, and Psychiatry* 20.1 (1957): 11.
Smith, C. N. and Squire L. R., 'Medial temporal lobe activity during retrieval of semantic memory is related to the age of the memory', *The Journal of Neuroscience* 29.4 (2009): 930–938.

Smith, E. E. and Jonides, J., 'Storage and executive processes in the frontal lobes', *Science* 283.5408 (1999): 1657–1661.

Villeda, S. A., et al., 'Young blood reverses age-related impairments in cognitive function and synaptic plasticity in mice', *Nature Medicine* 20, no. 6 (2014): 659–663.

The Brain's GPS

Dar-Nimrod, I. and Heine, S. J., 'Exposure to scientific theories affects women's math performance', *Science* 314, no. 5798 (2006): 435–435.

Hafting, T., et al., 'Microstructure of a spatial map in the entorhinal cortex', *Nature* 436, no. 7052 (2005): 801–806.

Ishikawa, T., et al., 'Wayfinding with a GPS-based mobile navigation system: A comparison with maps and direct experience', *Journal of Environmental Psychology* 28, no. 1 (2008): 74–82.

Jacobs, J., et al., 'Direct recordings of grid-like neuronal activity in human spatial navigation', *Nature Neuroscience* 16, no. 9 (2013): 1188–1190.

Jankowski, M. M., et al., 'The anterior thalamus provides a subcortical circuit supporting memory and spatial navigation', *Frontiers in Systems Neuroscience* 7 (2013).

Jog, M. S., et al., 'Building neural representations of habits', *Science* 286, no. 5445 (1999): 1745–1749.

Konishi K. and Bohbot V. D., 'Grey matter in the hippocampus correlates with spatial memory strategies in human older adults tested on a virtual navigation task', *Abstract Society for Neuroscience's annual meeting* (2010).

Kropff, E., et al., 'Speed cells in the medial entorhinal cortex', *Nature* (2015).

Maguire, E. A., et al., 'Navigation-related structural change in the hippocampi of taxi drivers', *Proceedings of the National Academy of Sciences* 97, no. 8 (2000): 4398–4403.

O'Keefe, J. and Dostrovsky, J., 'The hippocampus as a spatial map: Preliminary evidence from unit activity in the freely-moving rat', *Brain research* 34.1 (1971): 171–175.

Pacheco-Cobos, L., et al., 'Sex differences in mushroom gathering: men expend more energy to obtain equivalent benefits', *Evolution and Human Behavior* 31, no. 4 (2010): 289–297.
Save, E., et al., 'Dissociation of the effects of bilateral lesions of the dorsal hippocampus and parietal cortex on path integration in the rat', *Behavioral neuroscience* 115, no. 6 (2001): 1212.
Solstad, T., et al., 'Representation of Geometric Borders in the Entorhinal Cortex', *J. Cutan. Pathol* 34 (2007): 7.
Takahashi, N. M., et al., 'Pure topographic disorientation due to right retrosplenial lesion', *Neurology* 49, no. 2 (1997): 464–469.
Woollett, K. and Maguire, E. A., 'Acquiring "the Knowledge" of London's layout drives structural brain changes', *Current Biology* 21, no. 24 (2011): 2109–2114.

The emotional brain

Adelmann, P. K. and Zajonc, R. B., 'Facial efference and the experience of emotion', *Annual Review of Psychology* 40, no. 1 (1989): 249–280.
Als, H., et al. 'Early experience alters brain function and structure', *Pediatrics, 113*(4), (2004): 846–857.
Bardgett, M. E., et al., 'Dopamine modulates effort-based decision making in rats', *Behavioral Neuroscience* 123, no. 2 (2009): 242.
Bick, J., et al., 'Effect of early institutionalization and foster care on longterm white matter development: a randomized clinical trial', *JAMA Pediatrics* 169, no. 3 (2015): 211–219.
Denson, T. F., et al., 'The angry brain: Neural correlates of anger, angry rumination, and aggressive personality', *Journal of Cognitive Neuroscience* 21, no. 4 (2009): 734–744.
Dreyfuss, F. and Czaczkes, J. W., 'Blood cholesterol and uric acid of healthy medical students under the stress of an examination', *AMA Archives of Internal Medicine, 103*(5), (1959): 708–711.

Finzi, E. and Wasserman, E., 'Treatment of depression with botulinum toxin A: a case series', *Dermatologic Surgery* 32, no. 5 (2006): 645–650.

Friedman, M., et al., 'Changes in the serum cholesterol and blood clotting time in men subjected to cyclic variation of occupational stress', *Circulation, 17*(5), (1958): 852–861.

Gan, J. O., et al., 'Dissociable cost and benefit encoding of future rewards by mesolimbic dopamine', *Nature Neuroscience* 13, no. 1 (2010): 25–27.

Gerhardt, S., 'Why love matters: How affection shapes a baby's brain', *Infant Observation* 9.3 (2006): 305–309.

Giltay, E. J., et al., 'Dispositional optimism and all-cause and cardiovascular mortality in a prospective cohort of elderly dutch men and women', *Archives of general psychiatry,* 61(11), (2004): 1126–1135.

Hennenlotter, A., et al., 'The link between facial feedback and neural activity within central circuitries of emotion – new insights from Botulinum toxin–induced denervation of frown muscles', *Cerebral Cortex* 19, no. 3 (2009): 537–542.

Kappes, A., et al., 'Mental contrasting instigates goal pursuit by linking obstacles of reality with instrumental behavior', *Journal of Experimental Social Psychology* 48, no. 4 (2012): 811–818.

Kool, W., et al., Neural and behavioral evidence for an intrinsic cost of self-control', *PloS one* 8, no. 8 (2013): e72626.

Laudenslager, M. L., et al., 'Coping and immunosuppression: Inescapable but not escapable shock suppresses lymphocyte proliferation', *Science,* 221(4610), (1983): 568–570.

Lemke, M. R., et al., 'Effects of the dopamine agonist pramipexole on depression, anhedonia and motor functioning in Parkinson's disease', *Journal of the Neurological Sciences* 248, no. 1 (2006): 266–270.

Luby, J. L., et al., 'Maternal support in early childhood predicts larger hippocampal volumes at school age', *Proceedings of the National Academy of Sciences,* 109(8), (2012): 2854–2859.

Lupien, S. J., et al., 'Cortisol levels during human aging

predict hippocampal atrophy and memory deficits', *Nature Neuroscience,* 1(1), (1998): 69–73.

Mann, J. J., 'Role of the serotonergic system in the pathogenesis of major depression and suicidal behavior', *Neuropsychopharmacology* 21 (1999): 99S–105S.

Maruta, T., et al., 'Optimists vs pessimists: survival rate among medical patients over a 30-year period', *Mayo Clinic Proceedings,* vol. 75, no. 2, Elsevier (2000): 140–143.

Nelson, C. A., et al., 'Cognitive recovery in socially deprived young children: The Bucharest Early Intervention Project', *Science,* 318 (5858), (2007): 1937–1940.

Radiolab. 'Blame', sesong 12, episode 2, [podcast] Tilgjengelig på: http://www.radiolab.org/story/317421-blame/

Remy, P., et al., 'Depression in Parkinson's disease: loss of dopamine and noradrenaline innervation in the limbic system', *Brain* 128, no. 6 (2005): 1314–1322.

Salamone, J. D., et al., 'Effort-related functions of nucleus accumbens dopamine and associated forebrain circuits', *Psychopharmacology* 191, no. 3 (2007): 461–482.

Schachter, S. and Singer, J., 'Cognitive, social, and physiological determinants of emotional state', *Psychological Review* 69, no. 5 (1962): 379.

Sell, A., et al., 'Formidability and the logic of human anger', *Proceedings of the National Academy of Sciences* 106, no. 35 (2009): 15073–15078.

Spitz, R. A., 'Emotional deprivation in infancy', [video]: https://www.youtube.com/watch?v=VvdOe10vrs4

Spitz, R. A. and Wolf, K. M., 'Anaclitic depression; an inquiry into the genesis of psychiatric conditions in early childhood, II', *The Psychoanalytic Study of the Child* (1946).

Stoléru, S., et al., 'Functional neuroimaging studies of sexual arousal and orgasm in healthy men and women: a review and meta-analysis', *Neuroscience & Biobehavioral Reviews* 36, no. 6 (2012): 1481–1509.

Ströhle, A., et al., 'Physical activity and prevalence and incidence of mental disorders in adolescents and young adults', *Psychological Medicine* 37, no. 11 (2007): 1657–1666.

Takahashi, H., et al., 'When your gain is my pain and your pain is my gain: neural correlates of envy and schadenfreude', *Science* 323, no. 5916 (2009): 937–939.

Treadway, M. T., et al., 'Dopaminergic mechanisms of individual differences in human effort-based decision-making', *The Journal of Neuroscience* 32, no. 18 (2012): 6170–6176.

Tye, K. M., et al., 'Dopamine neurons modulate neural encoding and expression of depression-related behaviour', *Nature* 493, no. 7433 (2013): 537–541.

Van Kleef, G. A., et al., 'The interpersonal effects of anger and happiness in negotiations', *Journal of Personality and Social Psychology* 86, no. 1 (2004): 57.

Wise, R. A. 'Dopamine, learning and motivation', *Nature Reviews Neuroscience* 5, no. 6 (2004): 483–494.

Intelligence

Andreasen, N. C., et al., 'Intelligence and brain structure in normal individuals', *American Journal of Psychiatry* 150 (1993): 130–134

Flynn, J. R., 'IQ gains over time: Toward finding the causes', *The Rising Curve: Long-term gains in IQ and related measures* (1998): 25–66.

Flynn, J. R., 'Searching for justice: the discovery of IQ gains over time', *American Psychologist* 54, no. 1 (1999): 5.

Gottfredson, L. S., 'Why g matters: The complexity of everyday life', *Intelligence* 24, no. 1 (1997): 79–132.

Kanazawa, S., 'Intelligence and physical attractiveness', *Intelligence* 39, no.1 (2011):7–14.

Neubauer, A. C., et al., 'Intelligence and neural efficiency: The influence of task content and sex on the brain-IQ relationship', *Intelligence* 30, no. 6 (2002): 515–536.

Raven, J. 'The Raven's progressive matrices: change and stability over culture and time', *Cognitive Psychology* 41, no. 1 (2000): 1–48.

Reiss, A. L., et al., 'Brain development, gender and IQ in children', *Brain* 119, no. 5 (1996): 1763–1774.

Sturlason, S. 'Håvamål', translation by Ludvig Holm-Olsen, Aschehoug (1993): 22.
Willerman, L. R., et al., 'In vivo brain size and intelligence', *Intelligence* 15, no. 2 (1991): 223–228.

Culture © the brain

Allen, K., and Blascovich, J., 'Effects of music on cardiovascular reactivity among surgeons', *Jama* 272.11 (1994): 882–884.
Baroncelli, L., et al., 'Nurturing brain plasticity: impact of environmental enrichment', *Cell Death & Differentiation* 17.7 (2010): 1092–1103.
Chabris, C. F., 'Prelude or requiem for the Mozart effect ?', *Nature* 400.6747 (1999): 826–827.
Fox, J. G. and Embrey, E. D., 'Music – an aid to productivity', *Applied Ergonomics* 3.4 (1972): 202–205.
Gallese, V. and Goldman, A., 'Mirror neurons and the simulation theory of mind-reading', *Trends in Cognitive Sciences* 2.12 (1998): 493–501.
Geertz, C. 'The interpretation of cultures: Selected essays', vol. 5019. *Basic Books*, 1973.
Hebb, D. O., 'The effects of early experience on problem solving at maturity', *American Psychologist* 2 (1947): 306–307.
Perham, N. and Vizard, J., 'Can preference for background music mediate the irrelevant sound effect?' *Applied Cognitive Psychology* 25.4 (2011): 625–631.
Rauscher, F. H., et al., 'Music and spatial task performance', *Nature* 365 (1993): 611.
Sale, A., et al., 'Environment and brain plasticity: towards an endogenous pharmacotherapy', *Physiological Reviews* 94.1 (2014): 189–234.
Salimpoor, V. N., et al., 'Anatomically distinct dopamine release during anticipation and experience of peak emotion to music', *Nature Neuroscience* 14.2 (2011): 257–262.
Tylor, E. B., *Primitive culture: researches into the development of mythology, philosophy, religion, art, and custom*. Vol. 1. Murray, 1871.

Eating with your brain

Agostoni, C., et al., 'Prolonged breast-feeding (six months or more) and milk fat content at six months are associated with higher developmental scores at one year of age within a breast-fed population', *Bioactive Components of Human Milk*, Springer US, 2001: 137–141.

Barson, J. R., et al., 'Positive relationship between dietary fat, ethanol intake, triglycerides, and hypothalamic peptides: counteraction by lipid-lowering drugs', *Alcohol* 43, no. 6 (2009): 433–441.

Bayol, S. A., et al., 'A maternal "junk food" diet in pregnancy and lactation promotes an exacerbated taste for "junk food" and a greater propensity for obesity in rat offspring', *British Journal of Nutrition* 98.04 (2007): 843–851.

Beauchamp, G. K. and Mennella, J. A., 'Early flavor learning and its impact on later feeding behavior', *Journal of Pediatric Gastroenterology and Nutrition* 48 (2009): S25–S30.

Blumenthal, D. M. and Gold, M. S., 'Neurobiology of food addiction', *Current Opinion in Clinical Nutrition & Metabolic Care* 13.4 (2010): 359–365.

Chang, G-Q, et al., 'Maternal high-fat diet and fetal programming: increased proliferation of hypothalamic peptide-producing neurons that increase risk for overeating and obesity', *The Journal of Neuroscience* 28, no. 46 (2008): 12107–12119.

Conquer, J. A., et al., 'Fatty acid analysis of blood plasma of patients with Alzheimer's disease, other types of dementia, and cognitive impairment', *Lipids* 35, no. 12 (2000): 1305–1312.

De Snoo, K., 'Das trinkende kind im uterus', *Gynecologic and Obstetric Investigation* 105.2-3 (1937): 88–97.

Geiger B. M., et al., 'Deficits of mesolimbic dopamine neurotransmission in rat dietary obesity', *Neuroscience.* (2009) 159: 1193–1199.

Glusman, G, et al., 'The complete human olfactory subgenome', *Genome research.* 11.5 (2001): 685–702.

Helland, I. B., et al., 'Maternal supplementation with very-long-chain n-3 fatty acids during pregnancy and lactation augments children's IQ at 4 years of age', *Pediatrics* 111, no. 1 (2003): e39–e44.

Kalmijn, S., 'Fatty acid intake and the risk of dementia and cognitive decline: a review of clinical and epidemiological studies', *The journal of nutrition, health & aging* 4.4 (1999): 202–207.

Liley, A. W. 'Disorders of amniotic fluid', *Pathophysiology of Gestation* 2 (1972): 157–206.

Mennella, J. A., et al., 'Garlic ingestion by pregnant women alters the odor of amniotic fluid', *Chemical Senses* 20.2 (1995): 207–209.

Mennella, J. A., et al., 'Prenatal and postnatal flavor learning by human infants', *Pediatrics* 107.6 (2001): e88–e88.

Moss, M. *Salt, sugar, fat: how the food giants hooked us,* Random House, 2013.

Suez, J., et al., 'Artificial sweeteners induce glucose intolerance by altering the gut microbiota', *Nature* 514, no. 7521 (2014): 181–186.

Sussman, D., et al., 'Effects of a ketogenic diet during pregnancy on embryonic growth in the mouse', *BMC Pregnancy and Childbirth* 13, no. 1 (2013): 1.

Tellez, L. A., et al., 'Glucose utilization rates regulate intake levels of artificial sweeteners', *The Journal of Physiology* 591, no. 22 (2013): 5727–5744.

Ventura, A. K. and Worobey, J., 'Early influences on the development of food preferences', *Current Biology* 23.9 (2013): R401–R408.

Xiang, M., et al., 'Long-chain polyunsaturated fatty acids in human milk and brain growth during early infancy', *Acta Paediatrica* 89, no. 2 (2000): 142–147.

Yang, Q., 'Gain weight by "going diet?", Artificial sweeteners and the neurobiology of sugar cravings: Neuroscience. *The Yale journal of biology and medicine* 83.2 (2010): 101.

Addiction

Arseneault, L., et al., 'Cannabis use in adolescence and risk for adult psychosis: longitudinal prospective study', *BMJ*, 325(7374), (2002): 1212–1213.

Chiriboga, C. A., 'Fetal alcohol and drug effects', *The Neurologist* 9.6 (2003): 267–279.

Dackis, C. A. and Gold, M. S., 'New concepts in cocaine addiction: the dopamine depletion hypothesis', *Neuroscience & Biobehavioral Reviews* 9, no. 3 (1985): 469–477.

Goldschmidt, L., et al., 'Effects of prenatal marijuana exposure on child behavior problems at age 10', *Neurotoxicology and Teratology* 22.3 (2000): 325–336.

Levin, E. D. and Rezvani, A. H., 'Nicotinic treatment for cognitive dysfunction', *Current Drug Targets-CNS & Neurological Disorders* 1.4 (2002): 423–431.

Li, W., et al., 'White matter impairment in chronic heroin dependence: a quantitative DTI study', *Brain Research*, 1531, (2013): 58–64.

Qiu, Y., et al., 'Progressive white matter microstructure damage in male chronic heroin dependent individuals: a DTI and TBSS study', *PloS One, 8*(5),(2013): e63212.

Quik, M., et al., 'Nicotine as a potential neuroprotective agent for Parkinson's disease', *Movement Disorders* 27.8 (2012): 947–957.

Richardson, G. A., et al., 'Prenatal alcohol and marijuana exposure: effects on neuropsychological outcomes at 10 years', *Neurotoxicology and Teratology, 24*(3), (2002): 309–320.

Roehrs, T. and Roth, T., 'Caffeine: sleep and daytime sleepiness', *Sleep Medicine Reviews* 12, no. 2 (2008): 153–162.

Sim-Selley, L. J., 'Regulation of cannabinoid CB1 receptors in the central nervous system by chronic cannabinoids', *Critical Reviews™ in Neurobiology* 15.2 (2003).

Zammit, S., et al., 'Self-reported cannabis use as a risk factor for schizophrenia in Swedish conscripts of 1969: historical cohort study', *BMJ*, 325(7374), (2002): 1199.

Reality versus perception

Baron, R. A., 'Environmentally induced positive affect: Its impact on self-efficacy, task performance, negotiation, and conflict', *Journal of Applied Social Psychology* 20.5 (1990): 368–384.

Zampini, M. and Spence, C., 'The role of auditory cues in modulating the perceived crispness and staleness of potato chips', *Journal of sensory studies* 19.5 (2004): 347–363.

INDEX

Page references to illustrations are in *italics,*

D

E

F

J

K

L

M

R

S

T